WJEC LEVEL 2
ADDITIONAL MATHEMATICS

Series editor Linda Mason
Author Andrew Ginty

Photo credits

p. 1 © Archivist / stock.adobe.com; **p. 23** © loops7 / Getty Images; **p. 40** © efks / stock.adobe.com; **p. 59** © Africa Studio / stock.adobe.com; **p. 84** © Graham Moore /123RF.com; **p. 96** © taxiberlin / stock.adobe.com; **p. 98** © Igor Groshev / stock.adobe.com; **p. 122** © pixel. / stock.adobe.com; **p. 155** © Bastos - Fotolia.com; **p. 181** © MinDof / stock.adobe.com

Every effort has been made to trace all copyright holders, but if any have been inadvertently overlooked, the Publishers will be pleased to make the necessary arrangements at the first opportunity.

Although every effort has been made to ensure that website addresses are correct at time of going to press, Hodder Education cannot be held responsible for the content of any website mentioned in this book. It is sometimes possible to find a relocated web page by typing in the address of the home page for a website in the URL window of your browser.

Hachette UK's policy is to use papers that are natural, renewable and recyclable products and made from wood grown in well-managed forests and other controlled sources. The logging and manufacturing processes are expected to conform to the environmental regulations of the country of origin.

Orders: please contact Hachette UK Distribution, Hely Hutchinson Centre, Milton Road, Didcot, Oxfordshire, OX11 7HH. Email education@hachette.co.uk. Telephone: +44 (0)1235 827827. Lines are open from 9 a.m. to 5 p.m., Monday to Friday. You can also order through our website: www.hoddereducation.co.uk

The authorised representative in the EEA is Hachette Ireland, 8 Castlecourt Centre, Dublin 15, D15 XTP3, Ireland (email: info@hbgi.ie)

ISBN: 978 1 3983 5932 1

© Andrew Ginty, Linda Mason 2022

First published in 2022 by

Hodder Education,
An Hachette UK Company
Carmelite House
50 Victoria Embankment
London EC4Y 0DZ
www.hoddereducation.co.uk

Impression number   10 9 8 7 6 5 4 3 2
Year   2026 2025 2024

All rights reserved. Apart from any use permitted under UK copyright law, no part of this publication may be reproduced or transmitted in any form or by any means, electronic or mechanical, including photocopying and recording, or held within any information storage and retrieval system, without permission in writing from the publisher or under licence from the Copyright Licensing Agency Limited. Further details of such licences (for reprographic reproduction) may be obtained from the Copyright Licensing Agency Limited, www.cla.co.uk

Cover photo © robin_ph/stock.adobe.com

Typeset in Bembo Std, 11/13 pts. by Aptara, Inc.

Printed and bound by CPI Group (UK) Ltd, Croydon, CR0 4YY

A catalogue record for this title is available from the British Library.

Contents

Introduction ... v

SECTION 1 ALGEBRA

1 Number and algebra I ... 1
1. Numbers and the number system ... 1
2. Simplifying expressions ... 4
3. Solving linear equations ... 7
4. Algebra and number ... 10
5. Expanding brackets ... 12
6. Manipulating surds ... 14
7. Understanding and using indices with negative and fractional values ... 18

2 Algebra II ... 23
1. Factorising ... 23
2. Rearranging formulae ... 28
3. Simplifying algebraic fractions ... 30
4. Solving linear equations involving fractions ... 34
5. Completing the square ... 35
6. The minimum value of a quadratic expression ... 37

3 Algebra III ... 40
1. Function notation ... 40
2. Graphs of functions ... 42
3. Graphs of linear functions ... 43
4. Finding the equation of a line ... 48
5. Graphs of quadratic functions ... 53

4 Algebra IV ... 59
1. Quadratic equations ... 59
2. Simultaneous equations in two unknowns ... 67
3. The remainder theorem ... 73
4. The factor theorem ... 74
5. Algebraic proof ... 80

SECTION 2 GEOMETRY

5 Coordinate geometry ... 84
1. Parallel and perpendicular lines ... 84
2. The distance between two points ... 85
3. The midpoint of a line joining two points ... 86
4. Equation of a straight line ... 90
5. The intersection of two lines ... 93

6 Geometry I ... 98
1. Pythagoras' theorem ... 98
2. Trigonometry in two dimensions ... 99
3. Trigonometric functions for angles of any size ... 107
4. The sine and cosine graphs ... 108
5. The tangent graph ... 110
6. Solution of trigonometric equations ... 110
7. $a \sin kx$, $a \cos kx$ and $a \tan kx$ ... 114

7 Geometry II ... 122
1. The area of a triangle ... 122
2. Arcs and sectors ... 126
3. The sine rule ... 128
4. The cosine rule ... 132
5. Using the sine and cosine rules together ... 135
6. Mensuration ... 138
7. Problems in three dimensions ... 142
8. Lines and planes in three dimensions ... 143

SECTION 3 CALCULUS

8 Calculus ... 155
1. The gradient of a curve ... 155
2. Differentiation ... 156
3. Differentiation using standard results ... 160

4	Tangents	165
5	Increasing and decreasing functions	169
6	The second derivative	170
7	Stationary points	172

9 Integration 181

1	The rule for integrating x^n where n is a positive integer	182
2	The integral notation	185
3	Definite integrals	187
4	Areas between a curve and the x-axis	189
5	Areas below the x-axis	192
6	The area between two curves	197

PRACTICE QUESTIONS

Practice questions Paper 1	201
Practice questions Paper 2	204

Key terms	207
Index	208

Introduction

This book has been written to support the WJEC Level 2 Certificate in Additional Mathematics, but you may also use it independently as an introduction to Mathematics beyond GCSE. It is expected that many of the students using this book could be working with little day-today teacher support, and with this in mind, the text has been written in an interactive way and the answers are fuller than is often the case in books of this nature.

The qualification is designed for high-achieving students who have already acquired, or are expected to achieve, grades B to A★ in GCSE Mathematics. It is hoped that many of these students will progress to study Mathematics at Advanced Level and beyond.

Higher-order mathematical skills are studied in greater depth with an emphasis on algebraic reasoning, rigorous argument and problem-solving skills. Students following this course will be well prepared to tackle a Level 3 Mathematics qualification.

The content is split into algebra, geometry (including mensuration and trigonometry) and calculus with each section containing work that stretches and challenges, and which goes beyond the GCSE Programme of Study. The topics are frequently linked together as progress is made through the book, highlighting the beauty and inter-connectedness of mathematics.

Each chapter begins with a quote, designed to engage and bring the topic to life and/or provide an alternative viewpoint. The chapters are then broken down into sub-sections, each with a short introduction followed by a number of worked examples (with solutions) covering important techniques and question styles, and finally, one or more sets of exercise questions. Coloured boxes, hints and notes help to clarify some of the key points.

In addition, each chapter includes a number of activities. These are often used to introduce a new concept, or to reinforce the examples in the text. Throughout the book, the emphasis is on understanding the mathematics being used, rather than merely being able to perform the calculations, but the exercises do, nonetheless, provide plenty of scope for practising basic techniques.

Four symbols are used throughout the book:

! This 'warning sign' alerts you either to restrictions that need to be imposed or to possible pitfalls.

SA This indicates a question requiring selection and application of mathematical methods. These questions will sometimes involve more than one topic area.

IR This indicates a question requiring interpretation and reasoning to solve a mathematical problem. These questions will also sometimes involve more than one topic area.

RWC This indicates a question that relates to real-world contexts.

Numerous 'Discussion points' are used throughout as prompts to help you understand the theory that has been, or is about to be, introduced. Answers to these are also included.

'Prior knowledge' boxes highlight the GCSE Mathematics, or content earlier in the book, that you should be familiar with before you tackle a topic.

'Future uses' sections explain how the mathematics covered in a chapter can be used for further study, including at later points in the book, while 'Real-world contexts' explain the applications of the mathematics covered in each chapter. Also at the end of each chapter you will find a list of learning outcomes and key points.

A short glossary of key words is provided following the two practice question papers. Answers to these, all exercise questions, activities and discussion points are then given at the back of the book.

It is hoped that students who use this book will develop a fascination for mathematics, be inspired and challenged by the rigorous nature of the course and be able to appreciate the power of mathematics for its own sake, as well as a problem-solving tool.

1 Number and algebra I

The Book of Nature is written in the language of mathematics.
— Galileo Galilei

Prior knowledge
You are expected to be familiar with all number and ratio topics from GCSE.

1 Numbers and the number system

Number will be tested implicitly throughout the course. The following examples and questions provide practice of some of the basic number skills that may be needed.

Example 1.1 Simplify the ratio 3 kilometres : 840 metres.

Solution

$$\begin{aligned}
3\,\text{km} : 840\,\text{m} &= 3000\,\text{m} : 840\,\text{m} \\
&= 3000 : 840 \\
&= 300 : 84 \\
&= 100 : 28 \\
&= 25 : 7
\end{aligned}$$

Numbers and the number system

Example 1.2

Work out 43% of 5680.

Solution

43% of $5680 = 0.43 \times 5680$
$= 2442.4$

Example 1.3

Increase 540 by 17.5%.

Solution

$540 + 17.5\%$ of $540 = 540 \times 117.5\%$
$= 540 \times 1.175$
$= 634.5$

Example 1.4

Without using a calculator, work out $\frac{9}{10} - \frac{2}{5} \div \frac{6}{7}$.

Solution

$\frac{9}{10} - \frac{2}{5} \div \frac{6}{7} = \frac{9}{10} - \frac{2}{5} \times \frac{7}{6}$

$= \frac{9}{10} - \frac{2 \times 7}{5 \times 6}$

$= \frac{9}{10} - \frac{1 \times 7}{5 \times 3}$

$= \frac{9}{10} - \frac{7}{15}$

$= \frac{27}{30} - \frac{14}{30}$

$= \frac{13}{30}$

Example 1.5

Given the ratios $x : y = 5 : 3$ and $y : z = 4 : 7$, work out the ratio $x : z$ in its simplest form.

Solution

$x : y = 20 : 12$ and $y : z = 12 : 21$

so $x : y : z = 20 : 12 : 21$

so $x : z = 20 : 21$

Example 1.6 Work out, giving your answer to 3 significant figures, $\dfrac{3.76 \times 34}{78.4 \times 980}$.

Solution

$$\dfrac{3.76 \times 34}{78.4 \times 980} = 1.663890046 \times 10^{-3}$$

$$= 0.001663890046$$

$$= 0.00166 \ (3 \text{ s.f.})$$

Exercise 1A

Do not use a calculator for the questions marked ★.

① ABCD is a straight line (not drawn to scale). AB = 4 cm, AC = 10 cm, AD = 22 cm.

A　　　　B　　　　　C　　　　　　　D

Work out these ratios, giving your answers in their simplest form.

(i)　AC : AB　　　(ii)　AB : BC　　　(iii)　AD : AB

(iv)　BC : CD　　　(v)　BD : BC

> ❗ If a question involving money requires an answer to be given in pounds and pence, remember to give any non-integer answers to 2 decimal places.

★② Work out

(i)　60% of £115　　(ii)　$33\tfrac{1}{3}$% of 780　　(iii)　17.5% of 64 cm.

③ Work out

(i)　95% of 7540　　(ii)　$12\tfrac{1}{2}$% of 53.76　　(iii)　4.2% of £150.

★④ (i)　Increase 80 by 5%.　　(ii)　Increase £240 by 75%.

(iii)　Decrease £20 by 40%.　　(iv)　Decrease 36 by $66\tfrac{2}{3}$%.

⑤ (i)　Increase 650 by 14%.　　(ii)　Decrease 3250 by 3.5%.

(iii)　Decrease £3650 by 64%.　　(iv)　Increase £46.30 by $5\tfrac{1}{2}$%.

★⑥ Work out, giving your answers as fractions in their simplest form,

(i)　$\dfrac{3}{5} + \dfrac{2}{3} \times \dfrac{5}{6}$　　(ii)　$\left(\dfrac{1}{2}\right)^3 \div 4$　　(iii)　$3\dfrac{2}{5} - \dfrac{3}{4}$.

⑦ (i)　Work out, giving your answer to 3 significant figures,　$52.7 \div 4.93$

(ii)　Work out, giving your answer to 2 significant figures,　$5.9 - 0.53 \times 1.8$

(iii)　Work out, giving your answer to 1 significant figure,　$0.23 \times 0.14 + 0.09^2$

(iv)　Work out, giving your answer to 2 decimal places,　$\dfrac{19 + 36}{144 - 52}$.

⑧ A bag contains blue, green and white beads.

The ratio of blue beads to green beads is 4 : 3.

The ratio of green beads to white beads is 2 : 7.

Work out the smallest possible number of beads in the bag.

⑨ 55% of teachers in a school are female. The other 36 teachers are male. Work out the number of teachers in the school.

Prior knowledge

You are expected to be familiar with all aspects of GCSE algebra.

2 Simplifying expressions

When you are asked to *simplify* an algebraic expression you need to write it in its most compact form. This will involve techniques such as collecting like terms, removing brackets, factorising and finding a common denominator (if the expression includes fractions).

Example 1.7

Simplify this expression.

$3a + 4b - 2c + a - 3b - c$

Solution

Expression $= 3a + a + 4b - 3b - 2c - c$ (collecting like terms)

$= 4a + b - 3c$

Example 1.8

Simplify this expression.

$2(3x - 4y) - 3(x + 2y)$

⚠ A common error in questions like this is to forget to multiply all terms in the second bracket by -3.

Solution

Expression $= 6x - 8y - 3x - 6y$ (removing the brackets)

$= 3x - 14y$ $-3 \times 2y = -6y$

Example 1.9

Simplify this expression.

$3x^2yz \times 2xy^3$

Solution

Expression $= (3 \times 2) \times (x^2 \times x) \times (y \times y^3) \times z$ (collecting like terms)

$= 6x^3y^4z$

Example 1.10

Simplify this expression.

$\dfrac{12a^3b^2c^2}{8ab^5c}$

Solution

Divide the numerator and denominator by their highest common factor.

$\dfrac{12a^3b^2c^2}{8ab^5c} = \dfrac{3a^{3-1}c^{2-1}}{2b^{5-2}}$

$= \dfrac{3a^2c}{2b^3}$

Note

It is not necessary to include the intermediate step shown here.

Example 1.11

Factorise this expression.

$3a^2b + 6ab^2$

Discussion point
→ Explain what the word *factorise* means.

Solution

First write the highest common factor of the two terms, and then work on the contents of the brackets.

$3a^2b + 6ab^2 = 3ab(a + 2b)$

← $3a^2b = 3ab \times a$ and $6ab^2 = 3ab \times 2b$

Example 1.12

Simplify this expression.

$\dfrac{2x^2}{3yz} \div \dfrac{4xy^2}{5z^2}$

Solution

$$\text{Expression} = \dfrac{2x^2}{3yz} \times \dfrac{5z^2}{4xy^2}$$

$$= \dfrac{10x^2z^2}{12xy^3z}$$

$$= \dfrac{5xz}{6y^3}$$

Example 1.13

Write as a single fraction

$\dfrac{x}{4t} - \dfrac{2y}{5t} + \dfrac{z}{2t}.$

Solution

$$\dfrac{x}{4t} - \dfrac{2y}{5t} + \dfrac{z}{2t} = \dfrac{5x}{20t} - \dfrac{8y}{20t} + \dfrac{10z}{20t}$$

← $20t$ is the lowest common multiple of $4t$, $5t$ and $2t$.

$$= \dfrac{5x - 8y + 10z}{20t}$$

Exercise 1B

① Simplify the following expressions.

(i) $12a + 3b - 7c - 2a - 4b + 5c$
(ii) $4x - 5y + 3z + 2x + 2y - 7z$
(iii) $3(5x - y) + 4(x + 2y)$
(iv) $2(p + 5q) - (p - 4q)$
(v) $x(x + 3) - x(x - 2)$
(vi) $a(2a + 3) + 3(3a - 4)$
(vii) $3p(q - p) - 3q(p - q)$
(viii) $5f(g + 2h) - 5g(h - f)$

Simplifying expressions

② Factorise the following expressions by taking out the highest common factor.
- (i) $8 + 10x^2$
- (ii) $6ab + 8bc$
- (iii) $2a^2 + 4ab$
- (iv) $pq^3 + p^3q$
- (v) $3x^2y + 6xy^4$
- (vi) $6p^3q - 4p^2q^2 + 2pq^3$
- (vii) $15lm^2 - 9l^3m^3 + 12l^2m^4$
- (viii) $84a^5b^4 - 96a^4b^5$

③ Simplify the following expressions and factorise the results.
- (i) $4(3x + 2y) + 8(x - 3y)$
- (ii) $x(x - 2) - x(x - 8) + 6$
- (iii) $x(y + z) - y(x + z)$
- (iv) $p(2q - r) + r(p - 2q)$
- (v) $k(l + m + n) - km$
- (vi) $a(a - 2) - a(a + 4) + 2(a - 4)$
- (vii) $3x(x + y) - 3y(x - 2y)$
- (viii) $a(a - 2) - a(a - 4) + 8$

④ Simplify the following expressions as much as possible.
- (i) $2a^2b \times 5ab^3$
- (ii) $6p^3q \times 2q^3r$
- (iii) $lm \times mn \times np$
- (iv) $3r^3 \times 6s^2 \times 2rs$
- (v) $ab \times 2bc \times 4cd \times 8de$
- (vi) $3xy^2 \times 4yz^2 \times 5x^2z$
- (vii) $2ab^3 \times 6a^4 \times 7b^6$
- (viii) $6p^2q^3r \times 7pq^5r^4$

⑤ Simplify the following fractions as much as possible.
- (i) $\dfrac{4a^2b}{2ab}$
- (ii) $\dfrac{p^2}{q} \times \dfrac{q^2}{p}$
- (iii) $\dfrac{8a}{3b^2} \times \dfrac{6b^3}{4a^2}$
- (iv) $\dfrac{3ab}{2c^2} \times \dfrac{4cd}{6a^2}$
- (v) $\dfrac{8xy^3z^2}{12yz}$
- (vi) $\dfrac{3a^2}{9b^3} \div \dfrac{2a^4}{15b}$
- (vii) $\dfrac{5p^3q}{8rs^2} \div \dfrac{15pq^5}{28r^4}$

⑥ Write the following expressions as single fractions.
- (i) $\dfrac{2a}{3} + \dfrac{a}{4}$
- (ii) $\dfrac{2x}{5} - \dfrac{x}{2} + \dfrac{3x}{4}$
- (iii) $\dfrac{4p}{3} - \dfrac{3p}{4}$
- (iv) $\dfrac{2s}{5} - \dfrac{s}{3} + \dfrac{4s}{15}$
- (v) $\dfrac{3b}{8} - \dfrac{b}{6} + \dfrac{5b}{24}$
- (vi) $\dfrac{3a}{b} - \dfrac{2a}{3b}$
- (vii) $\dfrac{5}{2p} - \dfrac{3}{2q}$
- (viii) $\dfrac{2x}{3y} - \dfrac{3x}{2y}$

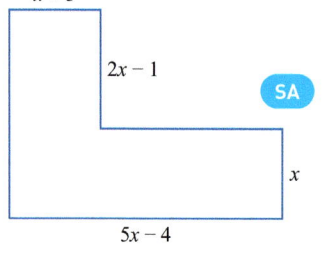

Figure 1.1

⑦ The angles of the hexagon in Figure 1.1 are all 90° or 270°.
Its side lengths are given in terms of x.
- (i) Work out its perimeter in terms of x.
- (ii) Work out its area in terms of x.

Give your answers in simplified form.

Figure 1.2

⑧ The rectangle in Figure 1.2 has length $5x + 2$ and width $3x - 1$.
Squares of side x are removed from each corner of the rectangle.
- (i) Write down a simplified expression for the perimeter of the new shape.
- (ii) Write down a simplified expression for the area of the new shape.

The new shape is the net of an open cuboid.
- (iii) Write down an expression for the volume of the cuboid.

3 Solving linear equations

Discussion points
→ What is an equation?
→ What does solving an equation mean?

Linear equations can be solved in different ways. The final answer will always be the same, regardless of the method used. The only rule to remember is that whatever is done to one side is also done to the other side. The following examples illustrate this. With practice you would probably omit some of the working.

Prior knowledge
You are expected to be familiar with solving linear equations.

Example 1.14

Solve this equation.

$3(3x - 17) = 2(x - 1)$

Solution

Multiply out the brackets	$\Rightarrow$	$9x - 51 = 2x - 2$
Subtract $2x$ from both sides	$\Rightarrow$	$9x - 51 - 2x = 2x - 2 - 2x$
Tidy up	$\Rightarrow$	$7x - 51 = -2$
Add 51 to both sides	$\Rightarrow$	$7x - 51 + 51 = -2 + 51$
Tidy up	$\Rightarrow$	$7x = 49$
Divide both sides by 7	$\Rightarrow$	$x = 7$

Example 1.15

Solve this equation.

$\frac{1}{2}(x + 8) = 2x + \frac{1}{3}(4x - 5)$

Solution

Start by clearing the fractions by multiplying both sides by 6 (the least common multiple of 2 and 3).

Discussion point
→ Why have the letter and the number swapped sides on the last line?

Multiply both sides by 6	$\Rightarrow$	$6 \times \frac{1}{2}(x + 8) = 6 \times 2x + 6 \times \frac{1}{3}(4x - 5)$
Tidy up	$\Rightarrow$	$3(x + 8) = 12x + 2(4x - 5)$
Multiply out the brackets	$\Rightarrow$	$3x + 24 = 12x + 8x - 10$
Tidy up	$\Rightarrow$	$3x + 24 = 20x - 10$
Subtract $3x$ from both sides	$\Rightarrow$	$24 = 17x - 10$
Add 10 to both sides	$\Rightarrow$	$34 = 17x$
Divide both sides by 17	$\Rightarrow$	$x = 2$

Solving linear equations

Sometimes you will need to set up the equation as well as solve it. When you are doing this, make sure that you define any variables you introduce.

Example 1.16

In a triangle, the largest angle is nine times as big as the smallest. The third angle is 60°.

(i) Write this information in the form of an equation.

(ii) Solve the equation to work out the sizes of the three angles.

Solution

(i) Let s = the smallest angle in degrees.

So $9s$ = the largest angle.

The sum of all three angles is 180°.

$s + 9s + 60 = 180$

(ii) Solving $\Rightarrow 10s = 120$.

$\Rightarrow s = 12$

The largest angle is then $9 \times 12 = 108$.

So the angles are 12°, 60° and 108°.

Exercise 1C

① Solve the following equations.

(i) $2x - 3 = x + 4$
(ii) $5a + 3 = 2a - 3$
(iii) $2(x + 5) = 14$
(iv) $7(2y - 5) = -7$
(v) $5(2c - 8) = 2(3c - 10)$
(vi) $3(p + 2) = 4(p - 1)$
(vii) $3(2x - 1) = 6(x + 2) + 3x$
(viii) $\frac{x}{3} + 7 = 5$
(ix) $\frac{5y - 2}{11} = 3$
(x) $\frac{k}{2} + \frac{k}{3} = 35$
(xi) $\frac{2t}{3} - \frac{3t}{5} = 4$
(xii) $\frac{5p - 4}{6} - \frac{2p + 3}{2} = 7$
(xiii) $p + \frac{1}{3}(p + 1) + \frac{1}{4}(p + 2) = \frac{5}{6}$

② The length, l metres, of a field is 80 m greater than the width. The perimeter is 600 m.

(i) Write this information in the form of an equation in l.

(ii) Work out the area of the field.

③ Rhys and Iwan are twins and their brother Steffan is four years younger. The total of their three ages is 17 years.

(i) Write this information in the form of an equation in s, Steffan's age in years.

(ii) What are all their ages?

SA ④ In a multiple-choice examination of 20 questions, four marks are given for each correct answer and one mark is deducted for each wrong answer. There is no penalty for not attempting a question. A candidate attempts a questions and gets c correct.

(i) Write down, and simplify, an expression for the candidate's total mark in terms of a and c.

(ii) A candidate attempts three-quarters of the questions and scores 40.
Write down, and solve, an equation for the number of correct answers.

SA ⑤ Marc is three times as old as his son, Jac, and in 12 years' time he will be twice as old as him.

(i) Given that Jac is j years old now, write an expression for Marc's age in 12 years' time.

(ii) Write down, and solve, an equation in j.

SA ⑥ A square has sides of length $2a$ metres, and a rectangle has length $3a$ metres and width 3 metres.

(i) Write down, in terms of a, the perimeter of the square.

(ii) Write down, in terms of a, the perimeter of the rectangle.

(iii) The perimeters of the square and the rectangle are equal.
Work out the value of a.

SA ⑦ The sum of five consecutive numbers is equal to 105. Let m represent the middle number.

(i) Write down the five numbers in terms of m.

(ii) Form an equation in m and solve it.

(iii) What are the five consecutive numbers?

IR ⑧ One rectangle has a length of $(x + 2)$ cm and a breadth of 2 cm. Another rectangle, of equal area, has a length of 5 cm and a width of $(x - 3)$ cm.
What is the area of each of the rectangles?

Discussion point

→ A large ice cream costs 40p more than a small one. Two large ice creams cost the same as three small ones. What is the cost of each size of ice cream?

→ This is an example of the type of question that you might find in a puzzle book or the puzzle section of a newspaper or magazine. How would you set about solving it?

Discussion point

→ You may think that the following question appears to be very similar to the one on the left. What happens when you try to solve it?

→ A large ice cream costs 40p more than a small one. Five small ice creams plus three large ones cost 80p less than three small ice creams plus five large ones. What is the cost of each size of ice cream?

Algebra and number

4 Algebra and number

Some algebra questions will involve using number skills.

Example 1.17

a is 75% of b and $b : c = 3 : 2$.

Show that $8a = 9c$.

Solution

a is 75% of b.

$a = \dfrac{75}{100}b$

$a = \dfrac{3}{4}b$ ①

$b : c = 3 : 2$

$\dfrac{b}{c} = \dfrac{3}{2}$

$b = \dfrac{3}{2}c$ ②

Substitute ② in ①.

$a = \dfrac{3}{4} \times \dfrac{3}{2}c$

$a = \dfrac{9}{8}c$

$8a = 9c$

Example 1.18

Write an expression for x increased by 13%.

Solution

x increased by 13% $= x + \dfrac{13}{100}x$

$= 1.13x$

Example 1.19

$p : q = 4 : 5$

Work out $p + 2q : 4q$, giving your answer in its simplest form.

Solution

Thinking in terms of parts:

p is 4 parts, q is 5 parts.

$p + 2q$ is $4 + 2 \times 5 = 14$ parts

$4q$ is 20 parts.

$p + 2q : 4q = 14 : 20$

$= 7 : 10$

An alternative solution is:

$$p : q = 4 : 5 \implies \frac{p}{q} = \frac{4}{5}$$

$$\implies p = \frac{4}{5}q$$

$$p + 2q : 4q = \frac{4}{5}q + 2q : 4q$$

$$= \frac{14}{5}q : 4q$$

$$= \frac{14}{5} : 4$$

$$= 7 : 10$$

Exercise 1D

1. Write expressions for the following, giving your answers in their simplest form.
   (i) 30% of b
   (ii) y% of 450
   (iii) c% of d

2. 60% of p = 40% of q.
   Work out p as a percentage of q.

3. Write expressions for the following, giving your answers in their simplest form.
   (i) a increased by 20%
   (ii) b increased by 5%
   (iii) k decreased by 35%
   (iv) m decreased by 2%

4. a increased by 80% is equal to b increased by 50%.
   Show that $\frac{b}{a} = 1.2$

5. (IR) p increased by 25% is equal to q decreased by 25%.
   Work out p as a percentage of q.

6. (IR) $x : y = 2 : 3$ and $y : z = 4 : 9$.
   Work out $x : y : z$, giving your answer in its simplest form.

7. (IR) $a : b = 5 : 2$
   (i) Write a in terms of b.
   (ii) Work out $2a + b : b$, giving your answer in its simplest form.
   (iii) Work out $7a - 5b : 4a$, giving your answer in its simplest form.

8. (IR) $m : n = 3 : 8$ and r is 20% of n.
   Work out $m : r$.

9. (IR) y is 20% greater than x.
   w is 20% less than y.
   Work out the ratio $w : x$ in its simplest form.

10. (IR) p is m% greater than q.
    p is m% less than r.
    Work out the ratio $r : q$ in terms of m.

11. (IR) The ratio of boys to girls in a room is 3 : 7.
    16 boys enter and 6 girls leave. The ratio is now 4 : 5.
    How many boys and how many girls are now in the room?

5 Expanding brackets

> **Prior knowledge**
>
> You are expected to be familiar with multiplication of two or three linear expressions.

> **Discussion point**
>
> → Why is $(x + 5)(2x - 3)$ a quadratic expression?

An expression of the form $ax^2 + bx + c$ (where the coefficient of x is non-zero) is a quadratic in x.

For example,

$x^2 + 3$

a^2 (a quadratic expression in a),

$2y^2 - 3y + 5$ (a quadratic expression in y).

Example 1.20

Expand $(x + 5)(2x - 3)$.

Solution

$$
\begin{aligned}
(x + 5)(2x - 3) &= x(2x - 3) + 5(2x - 3) \\
&= 2x^2 - 3x + 10x - 15 \\
&= 2x^2 + 7x - 15
\end{aligned}
$$

This method has multiplied everything in the second bracket by each term in the first bracket. An alternative way of setting this out is used in the next example.

Example 1.21

Expand $(3x - 5)^2$.

Solution

$(3x - 5)^2 = (3x - 5)(3x - 5)$ ← Write the square as the product of two brackets so you don't forget the middle term.

$$
\begin{array}{r}
3x - 5 \\
\times\ 3x - 5 \\
\hline
-15x + 25 \\
9x^2 - 15x \\
\hline
9x^2 - 30x + 25
\end{array}
$$

← Multiply the top line by -5.

← Multiply the top line by $3x$.

← Add the two products.

Example 1.22

Multiply $(x^3 + 2x - 4)$ by $(x^2 - x + 3)$.

Solution

$$(x^3 + 2x - 4)(x^2 - x + 3)$$
$$= x^3(x^2 - x + 3) + 2x(x^2 - x + 3) - 4(x^2 - x + 3)$$
$$= x^5 - x^4 + 3x^3 + 2x^3 - 2x^2 + 6x - 4x^2 + 4x - 12$$
$$= x^5 - x^4 + 5x^3 - 6x^2 + 10x - 12$$

Example 1.23

Expand and simplify $(a - 2)^3$.

Solution

$(a - 2)^3 = (a - 2)(a - 2)^2$

First, work out $(a - 2)^2$.

$$\begin{aligned}(a - 2)(a - 2) &= a(a - 2) - 2(a - 2) \\ &= a^2 - 2a - 2a + 4 \\ &= a^2 - 4a + 4\end{aligned}$$

Then multiply this by $(a - 2)$.

$$\begin{aligned}(a - 2)^3 &= (a - 2)(a^2 - 4a + 4) \\ &= a(a^2 - 4a + 4) - 2(a^2 - 4a + 4) \\ &= a^3 - 4a^2 + 4a - 2a^2 + 8a - 8 \\ &= a^3 - 6a^2 + 12a - 8\end{aligned}$$

Exercise 1E

① Expand the following expressions.
   (i) $(x + 5)(x + 4)$
   (ii) $(x + 3)(x + 1)$
   (iii) $(a + 5)(2a - 1)$
   (iv) $(2p + 3)(3p - 2)$
   (v) $(x + 3)^2$
   (vi) $(2x + 3)(2x - 3)$
   (vii) $(2 - 3m)(m - 4)$
   (viii)$(6 + 5t)(2 - t)$
   (ix) $(4 - 3x)^2$
   (x) $(m - 3n)^2$

② (i)   Multiply $(x^3 - x^2 + x - 2)$ by $(x^2 + 1)$.
   (ii) Multiply $(x^4 - 2x^2 + 3)$ by $(x^2 + 2x - 1)$.
   (iii) Multiply $(2x^3 - 3x + 5)$ by $(x^2 - 2x + 1)$.
   (iv) Multiply $(x^5 + x^4 + x^3 + x^2 + x + 1)$ by $(x - 1)$.
   (v)   Expand $(x + 2)(x - 1)(x + 3)$.

   Hint: Expand the first two sets of brackets first.

   (vi) Expand $(2x + 1)(x - 2)(x + 4)$.
   (vii) Expand and simplify $(x + 1)^3$.
   (viii)Expand and simplify $(p - 5)^3$.
   (ix) Expand and simplify $(2a + 3)^3$.
   (x)   Simplify $(2x^2 - 1)(x + 2) - 4(x + 2)^2$.
   (xi) Simplify $(x^2 - 1)(x + 1) - (x^2 + 1)(x - 1)$.

Manipulating surds

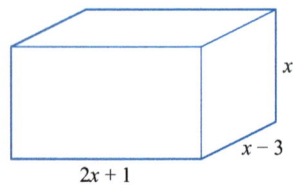

Figure 1.3

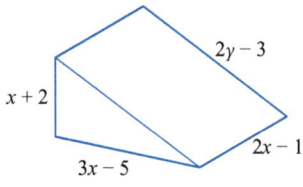

Figure 1.4

③ The cuboid in Figure 1.3 has length $(2x + 1)$, width $(x - 3)$, and height x.
  (i) Work out its volume.
  (ii) Work out its surface area.
  Leave your answers in expanded and simplified form.

④ The prism in Figure 1.4 has three rectangular faces, and two congruent right-angled triangular faces. It has length $(3x - 5)$ and height $(x + 2)$. Its slant height is $(2y - 3)$ and its width is $(2x - 1)$.
  (i) Work out its volume.
  (ii) Work out its surface area.
  Leave your answers in expanded and simplified form.

⑤ (i) Expand and simplify $(a + b)^2$.
  (ii) Hence expand and simplify $(a + b)^3$.

⑥ Use your answers to question 5 to write down the expansions of
  (i) $(x + 6)^3$
  (ii) $(p - 2)^3$.

> In the expansion of $(a + b)^3$ replace a with x, and b with 6.

Prior knowledge

You are expected to be familiar with the manipulation of surd expressions.

6 Manipulating surds

Simplifying expressions containing square roots

In mathematics there are times when it is helpful to be able to manipulate square roots, rather than just find their values from a calculator. This ensures that you are working with the exact value, not just a rounded version.

Example 1.24

Simplify the following.

(i) $\sqrt{8}$
(ii) $\sqrt{32} - \sqrt{18}$
(iii) $\sqrt{6} \times \sqrt{3}$
(iv) $(4 + \sqrt{3})(4 - \sqrt{3})$

Solution

(i) $\sqrt{8} = \sqrt{2 \times 2 \times 2}$
$= \sqrt{2} \times \sqrt{2} \times \sqrt{2}$
$= (\sqrt{2})^2 \times \sqrt{2}$
$= 2\sqrt{2}$

(ii) $\sqrt{6} \times \sqrt{3} = \sqrt{6 \times 3}$
$= \sqrt{2 \times 3 \times 3}$
$= (\sqrt{3})^2 \times \sqrt{2}$
$= 3\sqrt{2}$

(iii) $\sqrt{32} - \sqrt{18} = \sqrt{16 \times 2} - \sqrt{9 \times 2}$
$= 4\sqrt{2} - 3\sqrt{2}$
$= \sqrt{2}$

> Look for square factors of 32 and 18.
> 16 is the largest square factor of 32.
> 9 is the largest square factor of 18.

(iv) $(4 + \sqrt{3})(4 - \sqrt{3}) = 16 - 4\sqrt{3} + 4\sqrt{3} - (\sqrt{3})^2$
$= 16 - 3$
$= 13$

Discussion point
→ What is a rational number?

Notice that in part (iv) of Example 1.24 there is no square root in the answer. In the next example, all the numbers involve fractions with a square root as part of the denominator. It is easier to work with numbers if any square roots are only part of the numerator. Manipulating a number to that form is called *rationalising the denominator*.

When the numerator and the denominator of a fraction are multiplied by the same number, then the value of the fraction stays the same. For example, $\frac{3}{5} = \frac{3 \times 2}{5 \times 2} = \frac{6}{10}$.

We use this principle when rationalising a denominator. In the next examples, and question 3 of Exercise 1F, the denominators have only one term. In each case, multiply both the numerator and denominator by this number, and then simplify.

Example 1.25

Simplify the following by rationalising their denominators.

(i) $\frac{2}{\sqrt{3}}$ (ii) $\sqrt{\frac{3}{5}}$ (iii) $\sqrt{\frac{3}{8}}$

Solution

(i) $\frac{2}{\sqrt{3}} = \frac{2}{\sqrt{3}} \times \frac{\sqrt{3}}{\sqrt{3}}$
$= \frac{2\sqrt{3}}{(\sqrt{3})^2}$
$= \frac{2\sqrt{3}}{3}$

(ii) $\sqrt{\frac{3}{5}} = \frac{\sqrt{3}}{\sqrt{5}}$
$= \frac{\sqrt{3}}{\sqrt{5}} \times \frac{\sqrt{5}}{\sqrt{5}}$
$= \frac{\sqrt{3} \times \sqrt{5}}{(\sqrt{5})^2}$
$= \frac{\sqrt{15}}{5}$

(iii) $\sqrt{\frac{3}{8}} = \frac{\sqrt{3}}{\sqrt{8}}$
$= \frac{\sqrt{3}}{2\sqrt{2}}$
$= \frac{\sqrt{3}}{2\sqrt{2}} \times \frac{\sqrt{2}}{\sqrt{2}}$
$= \frac{\sqrt{3} \times \sqrt{2}}{2(\sqrt{2})^2}$
$= \frac{\sqrt{6}}{4}$

Manipulating surds

Exercise 1F

Do not use a calculator for this exercise.

① Simplify the following.

(i) $\sqrt{32}$ (ii) $\sqrt{125}$

(iii) $\sqrt{5} \times \sqrt{15}$ (iv) $\sqrt{8} - \sqrt{2}$

(v) $3\sqrt{27} - 6\sqrt{3}$ (vi) $4(3 + \sqrt{2}) - 3(5 - \sqrt{2})$

(vii) $4\sqrt{32} - 3\sqrt{8}$ (viii) $5(6 - \sqrt{3}) + 2(3 + 4\sqrt{3})$

(ix) $2\sqrt{125} + 6\sqrt{5}$ (x) $3(2\sqrt{2} - 3\sqrt{3}) - 2(3\sqrt{2} - 5\sqrt{3})$

② Simplify the following.

(i) $(\sqrt{2} - 1)^2$ (ii) $(4 - \sqrt{5})(2 + \sqrt{5})$

(iii) $(2 - \sqrt{7})(\sqrt{7} - 1)$ (iv) $(\sqrt{5} - \sqrt{3})(\sqrt{5} + \sqrt{3})$

(v) $(3 + \sqrt{2})(5 - 2\sqrt{2})$ (vi) $(\sqrt{7} - 3)(2\sqrt{7} + 3)$

(vii) $(3\sqrt{3} - 2)(2\sqrt{3} - 3)$ (viii) $(\sqrt{5} - \sqrt{3})^2$

(ix) $(5 - 3\sqrt{2})(2\sqrt{2} - 1)$ (x) $(2\sqrt{2} + 3)^2$

(xi) $(\sqrt{5} + \sqrt{3})^2 + (\sqrt{5} - \sqrt{3})^2$ (xii) $(\sqrt{7} + \sqrt{2})^2 - (\sqrt{7} - \sqrt{2})^2$

③ Simplify the following by rationalising their denominators.

(i) $\dfrac{1}{\sqrt{3}}$ (ii) $\dfrac{5}{\sqrt{5}}$ (iii) $\dfrac{8}{\sqrt{6}}$

(iv) $\sqrt{\dfrac{2}{3}}$ (v) $\dfrac{2\sqrt{2}}{\sqrt{8}}$ (vi) $\sqrt{\dfrac{3}{7}}$

(vii) $\dfrac{21}{\sqrt{7}}$ (viii) $\dfrac{5}{3\sqrt{5}}$ (ix) $\dfrac{\sqrt{75}}{\sqrt{125}}$

(x) $\dfrac{8}{\sqrt{128}}$

Hint: Remember that fractions can only be added or subtracted if they have common denominators.

④ Simplify the following by writing them as single fractions.

(i) $\dfrac{2}{3 - \sqrt{2}} + \dfrac{2}{3 + \sqrt{2}}$ (ii) $\dfrac{5}{2 - \sqrt{3}} - \dfrac{3}{2 + \sqrt{3}}$

(iii) $\dfrac{1}{5 - 2\sqrt{6}} + \dfrac{3}{5 + 2\sqrt{6}}$ (iv) $\dfrac{4}{4 + \sqrt{3}} - \dfrac{1}{4 - \sqrt{3}}$

⑤ Solve the following equations.

(i) $\sqrt{32} - v\sqrt{2} = \sqrt{8}$ (ii) $w\sqrt{18} + \sqrt{8} = \sqrt{98}$

(iii) $3\sqrt{3} + y\sqrt{12} = 2\sqrt{27}$ (iv) $x\sqrt{50} + \sqrt{18} = 5x\sqrt{8}$

⑥ Simplify $(2 + \sqrt{3})^3$

Hint: Simplify the surds and then multiply throughout each equation by the lowest common denominator.

⑦ Solve the following equations.

(i) $\dfrac{m}{\sqrt{3}} + \dfrac{1}{\sqrt{12}} = \sqrt{3}$ (ii) $\dfrac{3n}{\sqrt{2}} - \dfrac{n+4}{\sqrt{8}} = \sqrt{18}$

(iii) $\dfrac{2x}{\sqrt{5}} = \sqrt{20} + \dfrac{x}{\sqrt{45}}$ (iv) $3\left(\dfrac{x}{\sqrt{2}} + \sqrt{8}\right) = \dfrac{5x}{\sqrt{18}} + \dfrac{3x}{\sqrt{32}}$

⑧ Solve $(x - \sqrt{2})^3 = x^2(x - \sqrt{18}) + 3\sqrt{8}$.

IR ⑨ The area of a square is $(7 + 4\sqrt{3})$ cm².

The length (in centimetres) of each side of the square can be written in the form $m + n\sqrt{3}$, where m and n are integers.

Find the perimeter (in centimetres) of the square, writing your answer in the form $p + \sqrt{q}$, where p and q are integers.

Rationalising denominators with two terms

For a denominator with two terms, the multiplier we use is the denominator with one of its signs changed.

Example 1.26

Rationalise the denominator of
$$\frac{3\sqrt{2}}{4 - \sqrt{5}}.$$

Solution

$$\frac{3\sqrt{2}}{4 - \sqrt{5}} = \frac{3\sqrt{2}}{4 - \sqrt{5}} \times \frac{4 + \sqrt{5}}{4 + \sqrt{5}}$$

$$= \frac{12\sqrt{2} + 3\sqrt{2}\sqrt{5}}{16 + 4\sqrt{5} - 4\sqrt{5} - (\sqrt{5})^2}$$

$$= \frac{12\sqrt{2} + 3\sqrt{10}}{16 - 5}$$

$$= \frac{12\sqrt{2} + 3\sqrt{10}}{11}$$

Example 1.27

Write $\dfrac{2\sqrt{3} - 4}{3\sqrt{3} + 5}$ in the form $a + b\sqrt{3}$, where a and b are integers.

Solution

$$\frac{2\sqrt{3} - 4}{3\sqrt{3} + 5} = \frac{2\sqrt{3} - 4}{3\sqrt{3} + 5} \times \frac{3\sqrt{3} - 5}{3\sqrt{3} - 5}$$

$$= \frac{6(\sqrt{3})^2 - 10\sqrt{3} - 12\sqrt{3} + 20}{9(\sqrt{3})^2 - 15\sqrt{3} + 15\sqrt{3} - 25}$$

$$= \frac{18 - 22\sqrt{3} + 20}{27 - 25}$$

$$= \frac{38 - 22\sqrt{3}}{2}$$

$$= 19 - 11\sqrt{3}$$

Understanding and using indices with negative and fractional values

Exercise 1G

Do not use a calculator for this exercise.

1. Rationalise the denominators of the following fractions.

   (i) $\dfrac{2\sqrt{3}}{5+\sqrt{2}}$ (ii) $\dfrac{\sqrt{7}}{4-\sqrt{2}}$ (iii) $\dfrac{3\sqrt{3}}{\sqrt{3}+1}$

   (iv) $\dfrac{2+\sqrt{2}}{3-\sqrt{2}}$ (v) $\dfrac{\sqrt{7}-3}{1-\sqrt{7}}$ (vi) $\dfrac{10+\sqrt{3}}{\sqrt{3}+\sqrt{2}}$

2. Write $\dfrac{3\sqrt{2}+6}{\sqrt{2}-1}$ in the form $a+b\sqrt{2}$, where a and b are integers.

3. Write $\dfrac{2\sqrt{5}}{4\sqrt{5}+9}$ in the form $c\sqrt{5}+d$, where c and d are integers.

4. Write $\dfrac{1+\sqrt{3}}{3+2\sqrt{3}}$ in the form $p+\dfrac{q}{r}\sqrt{3}$, where p, q and r are integers.

5. **SA** A rectangle has a width of $(2+\sqrt{5})$ cm and an area of $(1+\sqrt{5})$ cm². Work out its length.

6. Simplify $\dfrac{19}{\sqrt{27}-\sqrt{8}}$.

7. **SA** A triangle has an area of $(11\sqrt{2}-2)$ cm² and a base length of $(2+3\sqrt{2})$ cm. Work out its perpendicular height.

8. **SA** The area of a trapezium is $(4+\sqrt{27})$ cm².
   The lengths of its parallel sides are $(3+\sqrt{12})$ cm and $(2-\sqrt{3})$ cm.
   Work out the perpendicular distance between the parallel sides.

7 Understanding and using indices with negative and fractional values

Useful index laws to learn:

- $a^m \times a^n = a^{m+n}$
- $a^m \div a^n = a^{m-n}$
- $(a^m)^n = a^{mn} = (a^n)^m$
- $a^{\frac{m}{n}} = \sqrt[n]{a^m} = \left(\sqrt[n]{a}\right)^m$
- $a^{-m} = \dfrac{1}{a^m}$
- $a^0 = 1$

Example 1.28

Simplify the following expressions.

(i) $a^3 \times a^4$ (ii) $c^8 \div c^5$ (iii) $(x^3)^5$

Solution

(i) $a^3 \times a^4 = a^7$ (ii) $c^8 \div c^5 = c^3$ (iii) $(x^3)^5 = x^{15}$

Example 1.29

Evaluate the following.

(i) $8^{\frac{4}{3}}$ (ii) 2^{-3} (iii) $27^{-\frac{2}{3}}$

Solution

(i) $8^{\frac{4}{3}} = \left(\sqrt[3]{8}\right)^4 = 2^4 = 16$

(ii) $2^{-3} = \dfrac{1}{2^3} = \dfrac{1}{8}$

(iii) $27^{-\frac{2}{3}} = 3^{-2} = \dfrac{1}{3^2} = \dfrac{1}{9}$

> Alternatively, the 4th power could be calculated first, followed by the cube root. But this would involve larger numbers:
> $8^{\frac{4}{3}} = \sqrt[3]{8^4} = \sqrt[3]{4096} = 16$

> The three operations can be carried out in any order. For example,
> $27^{-\frac{2}{3}} = \dfrac{1}{27^{\frac{2}{3}}} = \dfrac{1}{\sqrt[3]{729}} = \dfrac{1}{9}$

Example 1.30

Simplify the following expressions.

(i) $\left(2x \times x^{\frac{1}{3}} \times 8x^{-\frac{1}{2}}\right)^{\frac{3}{4}}$

(ii) $\dfrac{9m^{\frac{4}{5}} - 6m^{\frac{2}{5}}}{12m^{\frac{3}{5}}}$

Solution

(i) $\left(2x \times x^{\frac{1}{3}} \times 8x^{-\frac{1}{2}}\right)^{\frac{3}{4}} = \left(16x^{1+\frac{1}{3}-\frac{1}{2}}\right)^{\frac{3}{4}} = \left(16x^{\frac{5}{6}}\right)^{\frac{3}{4}} = 16^{\frac{3}{4}}\left(x^{\frac{5}{6}}\right)^{\frac{3}{4}}$

$= 2^3 x^{\frac{5}{6} \times \frac{3}{4}} = 8x^{\frac{5}{8}}$

(ii) $\dfrac{9m^{\frac{4}{5}} - 6m^{\frac{2}{5}}}{12m^{\frac{3}{5}}} = \dfrac{3m^{\frac{2}{5}}\left(3m^{\frac{2}{5}} - 2\right)}{12m^{\frac{3}{5}}} = \dfrac{3m^{\frac{2}{5}} - 2}{4m^{\frac{1}{5}}}$

> Factorise the numerator and then cancel any common factors with the denominator.

> Alternatively, the expression can be split into separate fractions:
> $\dfrac{9m^{\frac{4}{5}}}{12m^{\frac{3}{5}}} - \dfrac{6m^{\frac{2}{5}}}{12m^{\frac{3}{5}}} = \dfrac{3m^{\frac{1}{5}}}{4} - \dfrac{1}{2m^{\frac{1}{5}}} = \dfrac{3}{4}m^{\frac{1}{5}} - \dfrac{1}{2}m^{-\frac{1}{5}}$

Understanding and using indices with negative and fractional values

Exercise 1H

Do not use a calculator for this exercise.

① Simplify the following.

(i) $a^7 \times a^2$ (ii) $c^7 \div c^2$

(iii) $(e^3)^8$ (iv) $(2g^4)^3$

(v) $5h^4 \times 3h^{-6}$ (vi) $12k^6 \div 3k^{-8}$

(vii) $(2m)^3 \times (3m)^2$ (viii) $(n^3)^4 \times 6n^8 \div 2n^5$

(ix) $(2n^5 \times 3n^4)^2$ (x) $(3p^5 \div 6p^{-2})^{-8}$

② Evaluate the following.

(i) 6^0 (ii) 3^{-2}

(iii) 7^{-1} (iv) $9^{\frac{1}{2}}$

(v) 2^{-7} (vi) $27^{\frac{1}{3}}$

(vii) $16^{\frac{1}{4}}$ (viii) $\left(\dfrac{1}{32}\right)^{\frac{1}{5}}$

(ix) $64^{\frac{1}{2}}$ (x) $64^{\frac{1}{3}}$

(xi) $64^{\frac{1}{6}}$ (xii) $\left(\dfrac{1}{7}\right)^{-1}$

③ Evaluate the following.

(i) $4^{\frac{3}{2}}$ (ii) $\left(\dfrac{7}{8}\right)^0$

(iii) $\left(\dfrac{5}{9}\right)^{-1}$ (iv) $16^{\frac{3}{4}}$

(v) $25^{\frac{3}{2}}$ (vi) $27^{\frac{4}{3}}$

(vii) $16^{-\frac{1}{2}}$ (viii) $8^{-\frac{1}{3}}$

(ix) $\left(\dfrac{4}{9}\right)^{\frac{1}{2}}$ (x) $(-8)^{\frac{1}{3}}$

(xi) $\left(1\dfrac{7}{9}\right)^{\frac{1}{2}}$ (xii) $\left(-3\dfrac{3}{8}\right)^{\frac{1}{3}}$

④ Simplify the following.

(i) $(m^3n^2)^2 \times (m^8n)^3$ (ii) $(6ac^3)^2 \div (4a^4c^2)^3$

(iii) $\dfrac{1}{x^{-2}}$ (iv) $\dfrac{6}{9y^{-5}}$

(v) $(m^3 \times m^5 \times m^6)^{\frac{1}{2}}$ (vi) $n^{\frac{1}{2}} \times n^{\frac{1}{3}} \times n^{\frac{1}{4}}$

(vii) $\dfrac{x^2 \times x^5}{x^8}$ (viii) $\dfrac{x^2 + x^5}{x^8}$

⑤ Evaluate the following.

(i) $8^{-\frac{2}{3}}$ (ii) $4^{-\frac{3}{2}}$

(iii) $27^{-\frac{4}{3}}$ (iv) $\left(\dfrac{1}{125}\right)^{-\frac{2}{3}}$

(v) $64^{\frac{5}{6}}$ (vi) $36^{\frac{1}{2}} \times 8^{\frac{2}{3}}$

(vii) $\left(8^{\frac{1}{3}}\right)^2 \times \left(9^{\frac{1}{2}}\right)^{-2}$ (viii) $\left(7^{\frac{1}{3}}\right)^6$

(ix) $\left(\frac{9}{16}\right)^{-\frac{3}{2}} \times \left(\frac{64}{27}\right)^{-\frac{2}{3}}$ (x) $\left(2\frac{7}{9}\right)^{-\frac{1}{2}}$

6 Simplify the following.

(i) $\dfrac{3m^5 \times 5m^3}{6m^{13}}$ (ii) $\dfrac{6a^{\frac{1}{2}} \times 4a^{\frac{3}{2}}}{10a^{\frac{7}{2}}}$

(iii) $\dfrac{5c^{\frac{2}{3}} \times 4c^{\frac{5}{3}}}{6c^{\frac{4}{3}}}$ (iv) $\dfrac{7d^{\frac{3}{5}} \times 5d^{-\frac{4}{5}}}{14d^{\frac{7}{5}}}$

(v) $\dfrac{8e^{-\frac{5}{4}} \times 4e^{\frac{7}{4}}}{12e^{-\frac{3}{2}}}$ (vi) $\dfrac{9g^{-\frac{4}{7}} \times 5g^{-\frac{8}{7}}}{g^{\frac{3}{4}}}$

7 Simplify the following.

(i) $\dfrac{3m^5 + 5m^3}{6m^{13}}$ (ii) $\dfrac{6a^{\frac{1}{2}} - 4a^{\frac{3}{2}}}{10a^{\frac{7}{2}}}$

(iii) $\dfrac{5c^{\frac{2}{3}} + c^{\frac{5}{3}}}{c^{\frac{4}{3}}}$ (iv) $\dfrac{10d^{\frac{3}{5}} - 5d^{\frac{4}{5}} + 20d^{\frac{2}{5}}}{15d^{\frac{7}{5}}}$

(v) $\dfrac{8e^{\frac{5}{4}} + 4e^{\frac{7}{4}}}{12e^{\frac{3}{2}}}$ (vi) $\dfrac{9h^{-\frac{5}{6}} - 5h^{-\frac{7}{6}}}{h^{\frac{3}{4}}}$

8 Simplify the following.

(i) $\dfrac{4x^{\frac{5}{9}} \times 5x^{\frac{3}{9}}}{12x^{\frac{2}{9}}}$ (ii) $\dfrac{4y^{\frac{5}{9}} + 5y^{\frac{3}{9}}}{12y^{\frac{2}{9}}}$

(iii) $\left(w^{\frac{1}{2}} \times w^{\frac{2}{3}} \times w^{\frac{3}{4}}\right)^{\frac{1}{3}}$ (iv) $\dfrac{3a^{\frac{3}{4}} \times 6a^{\frac{1}{4}}}{9a^{\frac{1}{3}}}$

(v) $\dfrac{3e^{\frac{3}{4}} - 6e^{\frac{1}{4}}}{9e^{\frac{1}{3}}}$ (vi) $\left(7g^{\frac{1}{2}} \times 7g^{\frac{2}{3}} \times 49g^{-\frac{3}{4}}\right)^{-\frac{1}{4}}$

(vii) $\dfrac{2x^{\frac{1}{2}} \times 5x^{\frac{2}{3}} \times 8x^{\frac{3}{4}}}{20x^{\frac{1}{4}}}$

IR **9** Write $2^{\frac{1}{3}} + 2^{\frac{1}{3}}$ in the form 2^n.

IR **10** Find the value of $\left(36^{\frac{1}{3}} \times 4^{-\frac{4}{3}} \times 9^{\frac{2}{3}}\right)^{-\frac{1}{2}}$

SA **11** Write $\dfrac{1}{\left(\sqrt[3]{x^2}\right)^{-\frac{2}{5}}}$ in the form x^n.

SA **12** Write $3^{-\frac{3}{2}} + 3^{-\frac{1}{2}} + 3^{\frac{1}{2}} + 3^{\frac{3}{2}}$ as a simplified surd.

IR **13** Write $5^{-20} + 5^{-20} + 5^{-20} + 5^{-20} + 5^{-20}$ in the form 5^n.

Learning outcomes

LEARNING OUTCOMES

Now you have finished this chapter, you should be able to
- simplify an algebraic expression
- solve a linear equation
- solve percentage problems
- solve ratio problems
- work out the product of two (or more) algebraic expressions
- manipulate expressions involving surds, including
  - simplifying surds
  - adding/subtracting compatible surds
  - rationalising a denominator of the form $\sqrt{a}$
  - rationalising a denominator of the form $a + \sqrt{b}$
  - rationalising a denominator of the form $\sqrt{a} + \sqrt{b}$
- understand and use indices with negative and fractional values.

KEY POINTS

1. Useful index laws to learn:
   - $a^m \times a^n = a^{m+n}$
   - $a^m \div a^n = a^{m-n}$
   - $(a^m)^n = a^{mn} = (a^n)^m$
   - $a^{\frac{m}{n}} = \sqrt[n]{a^m} = \left(\sqrt[n]{a}\right)^m$
   - $a^{-m} = \dfrac{1}{a^m}$
   - $a^0 = 1$

2. Simplify algebraic expressions by collecting like terms and/or expanding brackets.
3. Add/subtract algebraic expressions by rewriting with common denominators.
4. Simplify fractions by cancelling common factors in the numerator and denominator.
5. When simplifying surds
   - only like surds can be added or subtracted
   - $\sqrt{a} \times \sqrt{b} = \sqrt{ab}$.
6. A denominator can be rationalised as follows:
   - $\sqrt{a}$ can be rationalised by using the multiplier $\sqrt{a}$
   - $a + \sqrt{b}$ can be rationalised by using the multiplier $a - \sqrt{b}$
   - a two-term expression can be rationalised using a multiplier found by changing one sign.

2 Algebra II

If A equals success, then the formula is A equals x plus y plus z, with x being work, y play, and z keeping your mouth shut.
Albert Einstein

Prior knowledge

You need to be able to identify a common factor of two or more terms. This may be just a number, a letter or both; see Chapter 1.

1 Factorising

Factorising an expression involves writing the expression as a product using brackets. Simple cases of this were seen in Chapter 1. Here, pairs of brackets will be needed. If you have already learnt another method, and use it quickly and accurately, then you should stick with it. With practice, you may be able to factorise some of these expressions **by inspection**.

You will meet factorising again in Chapter 4.

> ❗ When factorising a quadratic with no constant term, only one bracket is required. For example, $2x^2 - 8x = 2x(x - 4)$.

Factorising

Example 2.1

Factorise $xa + xb + ya + yb$.

Solution

First take out a common factor of each pair of terms.

$$\Rightarrow \quad xa + xb + ya + yb = x(a + b) + y(a + b)$$

Next notice that $(a + b)$ is now a common factor.

$$\Rightarrow \quad x(a + b) + y(a + b) = (a + b)(x + y)$$

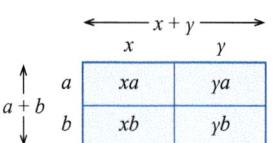

Figure 2.1

In practice this can relate to areas of rectangles.

The idea illustrated in Figure 2.1 can be used to factorise a quadratic expression containing three terms, but first you must decide how to split up the term in x.

Example 2.2

Factorise $x^2 + 6x + 8$.

Solution

Splitting the $6x$ as $4x + 2x$ gives

$$x^2 + 6x + 8 = x^2 + 4x + 2x + 8$$
$$= x(x + 4) + 2(x + 4)$$
$$= (x + 4)(x + 2).$$

Why would you choose to split up $6x$ this way?

Discussion point

→ Is the illustration in Figure 2.2 the only possibility?

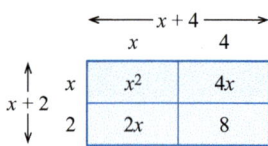

Figure 2.2

The crucial step is knowing how to split up the middle term.

To answer this question, notice that

- the numbers 4 and 2 have a sum of 6, which is the *coefficient* of x (i.e. the number multiplying x) in $x^2 + 6x + 8$
- the numbers 4 and 2 have a product of 8 which is the *constant term* in $x^2 + 6x + 8$.

There is only one pair of numbers that satisfies both of these conditions.

Example 2.3

Factorise $x^2 - 7x - 18$.

Solution

Pairs of numbers with a product of (-18) are:

1 and (-18)

2 and (-9)

3 and (-6)

6 and (-3)

9 and (-2)

18 and (-1).

Since the pair of numbers that you are looking for is unique, you can stop listing products when you find one that has the correct sum.

Discussion point

→ Do you get the same factors if the order in which you use the 2 and the (-9) is reversed so that you write it $x^2 - 9x + 2x - 18$?

There is only one pair, 2 and (−9), with a sum of (−7) so use these.

$$x^2 - 7x - 18 = x^2 + 2x - 9x - 18$$
$$= x(x + 2) - 9(x + 2)$$
$$= (x + 2)(x - 9)$$

Notice the sign change due to the − sign in front of the 9.

Example 2.4

Factorise $x^2 - 16$

Solution

First write $x^2 - 16 = x^2 + 0x - 16$.

Pairs of numbers with a product of (−16) are:

    1 and (−16)

    2 and (−8)

    4 and (−4),…(stop here).

The only pair with a sum of 0 is 4 and (−4).

$$x^2 - 16 = x^2 + 4x - 4x - 16$$
$$= x(x + 4) - 4(x + 4)$$
$$= (x + 4)(x - 4)$$

This is an example of a special case called the *difference of two squares* since you have $x^2 - 4^2 = (x + 4)(x - 4)$.

In general, $a^2 - b^2 = (a + b)(a - b)$.

Most people recognise this when it occurs and write down the answer straight away.

Example 2.5

Factorise $4x^2 - 9y^2$.

Solution

$$4x^2 - 9y^2 = (2x)^2 - (3y)^2$$
$$= (2x + 3y)(2x - 3y)$$

Notice that this technique can be extended to any situation where the coefficients of the two terms are square numbers.

Example 2.6

Factorise fully $y^5 - 36y^3$.

Solution

The instruction to factorise fully tells you that there is likely to be more than one step involved.

Take out the highest common factor of the two terms $y^5 - 36y^3 = y^3(y^2 - 36)$.

Use the difference of two squares $y^3(y^2 - 36) = y^3(y + 6)(y - 6)$.

Factorising

The technique for finding how to split the middle term needs modifying for examples where the expression starts with a multiple of x^2. The difference is that you now multiply the two outside numbers together to give the product you want.

Example 2.7

Factorise $2x^2 - 11x + 15$.

> A negative sum and a positive product means that both numbers are negative.

Solution

Here the sum is (-11) and the product is $2 \times 15 = 30$.

Options to give the correct product are:

(-1) and (-30) (-5) and (-6) ← (-5) and (-6), in either order, are the only options giving a sum of (-11).

(-2) and (-15)

(-3) and (-10).

$$2x^2 - 11x + 15 = 2x^2 - 5x - 6x + 15$$
$$= x(2x - 5) - 3(2x - 5)$$
$$= (x - 3)(2x - 5)$$

> **Discussion point**
> ➜ Try this example writing $2x^2 - 11x + 15$ as $2x^2 - 6x - 5x + 15$

Example 2.8

Factorise $3x^2 - 10xy - 8y^2$.

Solution

This expression can be factorised using the same method used in the previous example.

Here the sum is (-10) and the product is $3 \times -8 = -24$. ← A negative product means that one number is positive and the other is negative.

The option needed is (-12) and 2.

$$3x^2 - 10xy - 8y^2 = 3x^2 - 12xy + 2xy - 8y^2$$
$$= 3x(x - 4y) + 2y(x - 4y)$$
$$= (3x + 2y)(x - 4y)$$

Example 2.9

Factorise $(x + 3)^2 - 4y^2$.

Solution

This example uses the difference of two squares.

Writing the expression as $(x + 3)^2 - (2y)^2$ and factorising gives

$$[(x + 3) + (2y)][(x + 3) - 2y]$$
$$= (x + 3 + 2y)(x + 3 - 2y).$$

Exercise 2A

① Factorise the following expressions.
  (i) $ab - ac + db - dc$
  (ii) $2xy + 2x + wy + w$
  (iii) $2pq - 8p - 3rq + 12r$
  (iv) $5 - 5m - 2n + 2nm$

② Factorise the following expressions.
  (i) $x^2 + 5x + 6$
  (ii) $y^2 - 5y + 4$
  (iii) $m^2 - 8m + 16$
  (iv) $m^2 - 8m + 15$
  (v) $x^2 + 3x - 10$
  (vi) $a^2 + 20a + 96$
  (vii) $x^2 - x - 6$
  (viii) $y^2 - 16y + 48$
  (ix) $k^2 + 10k + 24$
  (x) $k^2 - 10k - 24$

③ Each of these is a difference of two squares. Factorise them.
  (i) $x^2 - 4$
  (ii) $a^2 - 25$
  (iii) $9 - p^2$
  (iv) $x^2 - y^2$
  (v) $t^2 - 64$
  (vi) $4x^2 - 1$
  (vii) $4x^2 - 9$
  (viii) $4x^2 - y^2$
  (ix) $16x^2 - 25$
  (x) $9a^2 - 4b^2$

④ Factorise the following expressions.
  (i) $2x^2 + 5x + 2$
  (ii) $2a^2 + 11a - 21$
  (iii) $15p^2 + 2p - 1$
  (iv) $3x^2 + 8x - 3$
  (v) $5a^2 - 9a - 2$
  (vi) $2p^2 + 5p - 3$
  (vii) $8x^2 + 10x - 3$
  (viii) $2a^2 - 3a - 27$
  (ix) $9x^2 - 30x + 25$
  (x) $4x^2 + 4x - 15$

> Example 2.8 shows you how to factorise an expression where there are two variables.

⑤ Factorise the following expressions.
  (i) $x^2 + 3xy + 2y^2$
  (ii) $x^2 + 4xy - 5y^2$
  (iii) $a^2 - ab - 12b^2$
  (iv) $c^2 - 11cd + 24d^2$
  (v) $x^2 + 9xy + 20y^2$
  (vi) $p^2 + 2pr - 15r^2$
  (vii) $a^2 - 2ar - 15r^2$
  (viii) $s^2 - 4st + 4t^2$
  (ix) $m^2 - 5mn - 6n^2$
  (x) $r^2 + 2rs - 8s^2$

⑥ Factorise the following expressions. (This question extends factorising using the difference of two squares as in Example 2.9.)
  (i) $(2a + 1)^2 - a^2$
  (ii) $(3x + 1)^2 - (x + 4)^2$
  (iii) $(2p - 3)^2 - (p + 1)^2$
  (iv) $16 - (5y - 2)^2$
  (v) $(2a + 1)^2 - a^2$
  (vi) $(3x + 1)^2 - (x + 4)^2$
  (vii) $(2p - 3)^2 - (p + 1)^2$
  (viii) $9 - (2y - 3)^2$

⑦ Factorise the following expressions.
  (i) $2x^2 + 5xy + 2y^2$
  (ii) $3x^2 + 5xy - 2y^2$
  (iii) $5a^2 - 8ab + 3b^2$
  (iv) $6c^2 + 5cd - 4d^2$
  (v) $6p^2 - 37pq + 6q^2$
  (vi) $7g^2 + 5gh - 2h^2$
  (vii) $6h^2 - 5hk - 4k^2$
  (viii) $8w^2 - 6wx + x^2$

⑧ Factorise fully the following expressions.
  (i) $x^3 - 4x$
  (ii) $a^4 - 16a^2$
  (iii) $9y^3 - y^5$
  (iv) $2x^3 - 2x$
  (v) $4p^4 - 9p^2$
  (vi) $100x - x^3$
  (vii) $18c^3 - 2c$
  (viii) $8x^3 - 50xy^2$

ACTIVITY 2.1

(i) Work out 9^2 and $(a^2)^2$. — Remember that $(a^p)^q = a^{pq}$.
(ii) Show that $a^4 - 81$ is the difference of two squares.
(iii) Factorise fully $a^4 - 81$.

Rearranging formulae

> **ACTIVITY 2.2**
> (i) Factorise $10x^2 + 11x + 3$.
> (ii) Factorise $10(p+q)^2 + 11(p+q) + 3$.

2 Rearranging formulae

C is called the subject of the formula.

The circumference of a circle is given by $C = 2\pi r$ where r is the radius. An equation such as this is often called a formula.

In some cases, you want to calculate r directly from C. You want r to be the subject of the formula.

Example 2.10

Make r the subject of $C = 2\pi r$.

Solution

Divide both sides by 2π $\Rightarrow \dfrac{C}{2\pi} = r$

$\Rightarrow r = \dfrac{C}{2\pi}$

> **!** Notice how the new subject should be on its own on the left-hand side of the new formula and must not appear on the right-hand side.

Example 2.11

Make x the subject of this formula.

$h = \sqrt{(x^2 + y^2)}$

A square root is assumed to be positive unless $\pm$ is added in front of it.

> **Discussion point**
> → What would you do with the $\pm$ sign in the case where h is the hypotenuse of a right-angled triangle with x and y as the other two sides?

Solution

Square both sides	$\Rightarrow$	$h^2 = x^2 + y^2$
Subtract y^2 from both sides	$\Rightarrow$	$h^2 - y^2 = x^2$
Make the x^2 term the subject	$\Rightarrow$	$x^2 = h^2 - y^2$
Take the square root of both sides	$\Rightarrow$	$x = \pm\sqrt{(h^2 - y^2)}$

Example 2.12

Make a the subject of this formula.

$v = u + at$

Solution

Subtract u from both sides	$\Rightarrow$	$v - u = at$
Divide both sides by t	$\Rightarrow$	$\dfrac{v-u}{t} = a$
Write the answer with a on the left-hand side	$\Rightarrow$	$a = \dfrac{v-u}{t}$

Exercise 2B

In this exercise all the equations refer to formulae used in mathematics. How many of them do you recognise?

① Make
  (i) u
  (ii) t
  the subject of $v = u + at$.

② Make b the subject of $A = \frac{1}{2}bh$.

③ Make l the subject of $P = 2(l + b)$.

④ Make r the subject of $A = \pi r^2$.

⑤ Make c the subject of $A = \frac{1}{2}(b + c)h$.

⑥ Make h the subject of $A = \pi r^2 + 2\pi rh$.

⑦ Make l the subject of $T = \frac{\lambda e}{l}$.

⑧ Make
  (i) u
  (ii) a
  the subject of $s = ut + \frac{1}{2}at^2$.

⑨ Make x the subject of $v^2 = \omega^2(a^2 - x^2)$.

The following examples show how to rearrange a formula when the letter that is to be the subject appears more than once.

Example 2.13

Make t the subject of this formula.

$at = 3(t + 2)$

Solution

Expand the brackets	$\Rightarrow$	$at = 3t + 6$
Collect all the terms in t on one side	$\Rightarrow$	$at - 3t = 6$
Factorise	$\Rightarrow$	$t(a - 3) = 6$
Divide both sides by $(a - 3)$	$\Rightarrow$	$t = \frac{6}{a - 3}$

The brackets are not needed in the denominator.

Example 2.14

Make x the subject of this formula.

$y = \frac{x + 2}{1 + 3x}$

Solution

Multiply both sides by $(1 + 3x)$	$\Rightarrow$	$y(1 + 3x) = x + 2$
Expand the brackets	$\Rightarrow$	$y + 3xy = x + 2$
Collect all the terms in x on one side and all the other terms on the other side	$\Rightarrow$	$3xy - x = 2 - y$
Factorise	$\Rightarrow$	$x(3y - 1) = 2 - y$
Divide both sides by $(3y - 1)$	$\Rightarrow$	$x = \frac{2 - y}{3y - 1}$

Simplifying algebraic fractions

Exercise 2C

① Make m the subject of $3m = x(m + 2)$.

② Make y the subject of $5y - 2x = xy$.

③ Make b the subject of $4(a + b) = 3(a - b)$.

④ Make h the subject of $S = 2\pi r^2 + 2\pi rh$.

⑤ Make x the subject of $y = \dfrac{x + 1}{2 + x}$.

⑥ Make c the subject of $d(2 + c) = 1 - 3c$.

⑦ (i) Make t the subject of $x = \dfrac{t}{t - 3}$.

   (ii) Hence, or otherwise, work out the value of t when $x = 3$.

⑧ (i) Make p the subject of $r = \dfrac{3p + 2}{2p + 3}$.

   (ii) Hence, or otherwise, work out the value of p when $r = -1$.

ACTIVITY 2.3

(i) (a) Show that $(x + 3)^2 = x^2 + 6x + 9$.

   (b) Hence make x the subject of $y = x^2 + 6x + 9$.

(ii) (a) Show that $(x - 5)^2 + 4 = x^2 - 10x + 29$.

   (b) Hence make x the subject of $p = x^2 - 10x + 29$.

3 Simplifying algebraic fractions

Prior knowledge

You should aim to be able to cancel fractions, find the least common multiple of two numbers and be able to identify a least common denominator of two or more fractions.

Discussion points

→ What is a fraction in arithmetic?
→ What about in algebra?

Fractions in algebra obey the same rules as fractions in arithmetic.

These cover two pairs of operations: × and ÷, and + and −.

Discussion points

→ When can you cancel fractions in arithmetic?
→ What about in algebra?
→ What is a factor in arithmetic?
→ What about in algebra?

Example 2.15

Simplify the following.

(i) $\dfrac{18}{24}$

(ii) $\dfrac{2x+2}{3x+3}$

(iii) $\dfrac{a^2 - a - 6}{a^2 - 8a + 15}$

Discussion points

→ Look at this calculation for (ii).
$$\dfrac{2\cancel{x}+2}{3\cancel{x}+3} = \dfrac{4}{6} = \dfrac{2}{3}$$
Why is it wrong?

→ Look at this calculation for (iii).
$$\dfrac{\cancel{a^2} - \cancel{a}^1 - 6}{_1\cancel{a^2} - 8\cancel{a}_1 + 15} = -\dfrac{6}{8} = -\dfrac{3}{4}$$
Why is it wrong?

Solution

(i) $\dfrac{18}{24} = \dfrac{^1\cancel{6} \times 3}{_1\cancel{6} \times 4} = \dfrac{3}{4}$

(ii) $\dfrac{2x+2}{3x+3} = \dfrac{2\,\cancel{(x+1)}^1}{3\,\cancel{(x+1)}_1} = \dfrac{2}{3}$

(iii) $\dfrac{a^2 - a - 6}{a^2 - 8a + 15} = \dfrac{\cancel{(a-3)}^1 (a+2)}{\cancel{(a-3)}_1 (a-5)} = \dfrac{a+2}{a-5}$

Example 2.16

Simplify the following.

(i) $\dfrac{2}{3} \times \dfrac{9}{14}$

(ii) $\dfrac{3}{4} \div \dfrac{9}{16}$

(iii) $\dfrac{3a^2 b}{2c} \times \dfrac{4c^3}{9ab}$

(iv) $\dfrac{4n^2 - 9}{n+1} \div \dfrac{2n+3}{n^2 - 1}$

Discussion point

→ Look at this calculation for (iv).
$$\dfrac{\cancel{4n^2}^{2n} - \cancel{9}^3}{\cancel{n}_1 + 1} \times \dfrac{\cancel{n^2}^n - 1}{2\cancel{n} + \cancel{3}_1}$$
$$= \dfrac{(2n-3)(n-1)}{4}$$
Why is it wrong?

Solution

(i) $\dfrac{^1\cancel{2}}{_1\cancel{3}} \times \dfrac{\cancel{9}^3}{\cancel{14}_7} = \dfrac{1 \times 3}{1 \times 7} = \dfrac{3}{7}$

(ii) $\dfrac{3}{4} \div \dfrac{9}{16} = \dfrac{\cancel{3}^1}{\cancel{4}_1} \times \dfrac{\cancel{16}^4}{\cancel{9}_3} = \dfrac{4}{3}$

(iii) $\dfrac{^1\cancel{3}a^2\cancel{b}}{_1\cancel{2c}} \times \dfrac{_2\cancel{4}c^{\cancel{3}^2}}{_3\cancel{9}\cancel{ab}} = \dfrac{2ac^2}{3}$

(iv) $\dfrac{4n^2 - 9}{n+1} \div \dfrac{2n+3}{n^2 - 1} = \dfrac{\cancel{(2n+3)}^1 (2n-3)}{\cancel{(n+1)}_1} \times \dfrac{\cancel{(n+1)}^1 (n-1)}{\cancel{(2n+3)}_1}$
$= (2n-3)(n-1)$

Example 2.17

Write the following as a single fraction and simplify where possible.

(i) $\dfrac{2}{3} + \dfrac{3}{4}$

(ii) $\dfrac{5x}{6} + \dfrac{x}{4}$

(iii) $\dfrac{2}{(x+1)} + \dfrac{5}{(x-1)}$

(iv) $\dfrac{a}{a^2 - 1} - \dfrac{2}{a+1}$

(v) $\dfrac{2}{x+3} - \dfrac{3}{x-1} + 4$

Simplifying algebraic fractions

Solution

(i) $\dfrac{2}{3} + \dfrac{3}{4} = \dfrac{8}{12} + \dfrac{9}{12} = \dfrac{17}{12}$

(ii) $\dfrac{5x}{6} + \dfrac{x}{4} = \dfrac{10x}{12} + \dfrac{3x}{12} = \dfrac{13x}{12}$

> Take care to ensure that the common denominator is the least common multiple of the original denominators.

(iii) $\dfrac{2}{(x+1)} + \dfrac{5}{(x-1)} = \dfrac{2(x-1)}{(x+1)(x-1)} + \dfrac{5(x+1)}{(x+1)(x-1)}$

$= \dfrac{2x - 2 + 5x + 5}{(x+1)(x-1)}$

$= \dfrac{7x + 3}{(x+1)(x-1)}$

(iv) $\dfrac{a}{a^2 - 1} - \dfrac{2}{a+1} = \dfrac{a}{(a-1)(a+1)} - \dfrac{2}{a+1}$

$= \dfrac{a}{(a-1)(a+1)} - \dfrac{2(a-1)}{(a-1)(a+1)}$

$= \dfrac{a - 2a + 2}{(a-1)(a+1)}$

$= \dfrac{2 - a}{(a-1)(a+1)}$

(v) $\dfrac{2}{x+3} - \dfrac{3}{x-1} + 4 = \dfrac{2}{x+3} - \dfrac{3}{x-1} + \dfrac{4}{1}$

$= \dfrac{2(x-1)}{(x+3)(x-1)} - \dfrac{3(x+3)}{(x-1)(x+3)} + \dfrac{4(x-1)(x+3)}{(x-1)(x+3)}$

$= \dfrac{2x - 2 - (3x + 9) + 4(x^2 + 2x - 3)}{(x-1)(x+3)}$

$= \dfrac{2x - 2 - 3x - 9 + 4x^2 + 8x - 12}{(x-1)(x+3)}$

$= \dfrac{4x^2 + 7x - 23}{(x-1)(x+3)}$

> The brackets in the numerator are expanded to allow the like terms to be collected and simplified. The denominator could be expanded to $x^2 + 2x - 3$ but is more useful if left factorised.

Exercise 2D

① Simplify the following.

(i) $\dfrac{2(x+3)}{4x+12}$ (ii) $\dfrac{4x - 8}{(x-2)(x+8)}$ (iii) $\dfrac{3(x+y)}{x^2 - y^2}$

(iv) $\dfrac{6x^2 y^3}{9xy^4}$ (v) $\dfrac{2p}{6p - 2p^2}$ (vi) $\dfrac{4ab^3}{10a^3 b}$

② Simplify the following.

(i) $\dfrac{x^2 - 4x + 3}{2x - 6}$ (ii) $\dfrac{x^2 + xy}{x^2 - y^2}$ (iii) $\dfrac{a+2}{a^2 - a - 6}$

(iv) $\dfrac{3x^2 + 15x}{10x + 2x^2}$ (v) $\dfrac{9x^2 - 1}{9x + 3}$ (vi) $\dfrac{3x^2 + 3xy}{6xy + 6y^2}$

③ Simplify the following.

(i) $\dfrac{3a}{b^2} \times \dfrac{b^3}{6a}$ (ii) $\dfrac{xy - y^2}{y} \times \dfrac{x}{x - y}$

(iii) $\dfrac{x^2 - y^2}{y} \times \dfrac{x}{x - y}$ (iv) $\dfrac{x + 1}{2x} \div \dfrac{4x^2 - 4}{x^2}$

(v) $\dfrac{3a^2 + a - 2}{2} \div \dfrac{6a^2 - a - 2}{8a + 4}$ (vi) $\dfrac{2p^2 - pq - q^2}{3p + 3q} \div \dfrac{2p^2 - 3pq + q^2}{2p + 2q}$

④ Simplify the following.

(i) $\dfrac{x^2 - 4x + 4}{x^2 - 2x} \times \dfrac{x - 2}{x^2 - 4}$ (ii) $\dfrac{2x - 1}{x + 1} \div \dfrac{2x^2 - x - 1}{x^2 + 3x + 2}$

(iii) $\dfrac{4p^2 + 12}{p - 3} \times \dfrac{p^2 - 9}{p^2 + 3}$ (iv) $\dfrac{3x^2 - 9}{x + 2} \div \dfrac{x^2 - 6x + 9}{x^2 + x - 2}$

(v) $\dfrac{3a^2 - 12}{5a^2 - 4a - 1} \times \dfrac{5a + 1}{(a - 2)^2}$ (vi) $\dfrac{2t}{t^2 + 1} \div \dfrac{4t^2}{t^4 - 1}$

⑤ Write the following as a single fraction and simplify where possible.

(i) $\dfrac{3a}{5} - \dfrac{a}{4}$ (ii) $\dfrac{5}{3a} - \dfrac{4}{a}$

(iii) $\dfrac{2}{(m + n)} - \dfrac{1}{(m - n)}$ (iv) $\dfrac{4}{p - 2} - \dfrac{3}{2p + 1}$

(v) $\dfrac{1}{2(x - 1)} + \dfrac{2}{(x + 4)}$ (vi) $\dfrac{1}{2(a - 1)} + \dfrac{2}{3(a + 4)}$

⑥ Write the following as a single fraction and simplify where possible.

(i) $\dfrac{2}{a^2 + a} + \dfrac{3}{a^2 - a}$ (ii) $\dfrac{2x}{x - y} + \dfrac{2y}{y - x}$

(iii) $\dfrac{p}{p^2 - 1} - \dfrac{1}{p + 1}$ (iv) $\dfrac{a - b}{a + b} + \dfrac{a + b}{a - b}$

(v) $\dfrac{4}{x^2 - 4} - \dfrac{3}{x + 2}$ (vi) $\dfrac{7}{5(x - 2)} - \dfrac{2}{x + 4}$

⑦ Write the following as a single fraction and simplify where possible.

(i) $\dfrac{1}{x + 1} - \dfrac{2}{x + 2} + \dfrac{3}{x + 3}$ (ii) $\dfrac{3}{x + 1} - \dfrac{2}{x - 2} + \dfrac{4}{x + 3}$

(iii) $\dfrac{x + 2}{(x + 1)^2} - \dfrac{1}{x}$

⑧ Write the following as a single fraction and simplify where possible.

(i) $\dfrac{4t}{t^2 + 2t + 1} + \dfrac{3}{t + 1}$ (ii) $\dfrac{1}{y^2 - x^2} + \dfrac{3}{y + x}$

(iii) $\dfrac{2}{x - 2} + \dfrac{1}{2x - 1} - 1$ (iv) $1 + \dfrac{1}{n} + \dfrac{1}{n + 1} + \dfrac{1}{n + 2}$

(v) $\dfrac{3}{x + 1} + \dfrac{2}{x - 1} - \dfrac{2}{3}$ (vi) $\dfrac{5}{2x + 1} - \dfrac{1}{3x - 2} + \dfrac{7}{8}$

4 Solving linear equations involving fractions

> **Prior knowledge**
>
> You are expected to be familiar with the basic rules for the mathematical operations $+$, $-$, $\times$ and $\div$ which are used in Chapter 1.

Example 2.18

Solve the following.

$$\frac{x+2}{6} = \frac{x-6}{2}$$

Solution

The LCM of 6 and 2 is 6, so multiply by 6.

$$^1\!\!\not{6} \times \frac{(x+2)}{\not{6}_1} = ^3\!\!\not{6} \times \frac{(x-6)}{\not{2}_1}$$

$$\Rightarrow x + 2 = 3x - 18$$
$$\Rightarrow 20 = 2x$$
$$\Rightarrow x = 10$$

> **Discussion point**
>
> → When you multiply a fraction by an integer, you only multiply its numerator (top line). Why?

Example 2.19

Solve the following.

$$\frac{x+2}{6} + 3 = \frac{x}{5}$$

Solution

The LCM of 6 and 5 is 30, so multiply by 30.

$$^5\!\!\not{30} \times \frac{(x+2)}{\not{6}_1} + 30 \times 3 = ^6\!\!\not{30} \times \frac{x}{\not{5}_1}$$

$$\Rightarrow 5x + 10 + 90 = 6x$$
$$\Rightarrow x = 100$$

> **Discussion point**
>
> → Look at this version of the first stage of the solution for Example 2.19.
>
> $$30 \times \frac{(x+2)}{6} + 3$$
> $$= 30 \times \frac{x}{5}$$
>
> Why is it wrong?

Exercise 2E

Solve the following equations.

① $x - \frac{x}{5} = \frac{2}{3}$

② $\frac{2}{a} - \frac{3}{4a} = 2$

③ $\frac{x-4}{6} = \frac{x+2}{3}$

④ $\frac{2-3x}{6} = \frac{2}{3}$

⑤ $\frac{3p+2}{2} - \frac{p-1}{5} = 3$

⑥ $\frac{3(x-2)}{2} - \frac{x-5}{4} = 2$

⑦ $x + 1 - \frac{3(x-2)}{2} = 7$

⑧ $\frac{3(t+4)}{8} + 2 = \frac{2t}{3}$

⑨ $\frac{x+1}{5} + \frac{2x-3}{6} = \frac{1}{3}$

⑩ $\frac{2x-1}{7} - \frac{x+3}{4} = \frac{2}{5}$

5 Completing the square

When considering a quadratic expression it will sometimes be useful to write it to include the term $(x + a)^2$ or $(x - a)^2$, where a is a constant. Some uses of this approach will be seen later in sections on quadratic equations and quadratic graphs.

You will meet completing the square again in Chapter 4.

Example 2.20

Work out the values of p and q such that $x^2 - 6x + 2 = (x - p)^2 + q$.

Solution

Expand the bracket $x^2 - 6x + 2 = x^2 - 2px + p^2 + q$

Equate coefficients of x $-6 = -2p$

$3 = p$

Equate constants $2 = p^2 + q$

$2 = 9 + q$

$-7 = q$

$p = 3$ and $q = -7$

'Equate coefficients of x' means making equal the number of x on each side of the identity.

$(x - p)^2 = (x - p)(x - p)$
$= x^2 - px - px + p^2$
$= x^2 - 2px + p^2$

Example 2.21

Work out the values of a, b and c such that $2x^2 + bx + 5 = a(x - 3)^2 + c$.

Solution

Expand the bracket $2x^2 + bx + 5 = a(x^2 - 6x + 9) + c$

$= ax^2 - 6ax + 9a + c$

Equate coefficients of x^2 $2 = a$

Equate coefficients of x $b = -6a$

$b = -12$

Equate constants $5 = 9a + c$

$5 = 18 + c$

$-13 = c$

$a = 2$, $b = -12$ and $c = -13$

Completing the square

Example 2.22

Work out the values of a, b and c such that $3x^2 + 5x - 1 = a(x+b)^2 + c$.

Solution

Expand the bracket
$$3x^2 + 5x - 1 = a(x^2 + 2bx + b^2) + c$$
$$= ax^2 + 2abx + ab^2 + c$$

Equate coefficients of x^2 $\quad 3 = a$

Equate coefficients of x $\quad 5 = 2ab$
$$5 = 6b$$
$$\frac{5}{6} = b$$

Equate constants $\quad -1 = ab^2 + c$
$$-1 = 3 \times \left(\frac{5}{6}\right)^2 + c$$
$$-1 = 3 \times \frac{25}{36} + c$$
$$-1 = \frac{25}{12} + c$$
$$-\frac{37}{12} = c$$

$$a = 3,\ b = \frac{5}{6}\ \text{and}\ c = -\frac{37}{12}$$

> **Note**
> Comparing coefficients is a useful technique which can be applied to any polynomial.
>
> An alternative method when rewriting a quadratic in the form $a(x+b)^2 + c$, is to use the 'completing the square' technique given in Chapter 4.

Exercise 2F

1. Work out the values of a and b such that $x^2 + 8x + 10 = (x+a)^2 + b$.
2. Work out the values of c and d such that $x^2 - cx + 7 = (x-1)^2 + d$.
3. Work out the values of p and q such that $x^2 - 12x - 4 = (x-p)^2 + q$.
4. Work out the values of a and b such that $x^2 + 5x - 2 = (x+a)^2 + b$.
5. Work out the values of p and q such that $5 + 4x - x^2 = p - (x-q)^2$.
6. Work out the values of c and d such that $2 - x - x^2 = c - (x+d)^2$.
7. Work out the values of a, b and c such that $2x^2 + bx + 5 = a(x+2)^2 + c$.
8. Work out the values of a, b and c such that $5x^2 + 30x + 10 = a(x+b)^2 + c$.
9. Work out the values of p, q and r such that $3x^2 - 12x + 14 = p(x+q)^2 + r$.
10. Work out the values of a, b and c such that $3x^2 - bx + 1 = a(x-4)^2 + c$.
11. Work out the values of a, b and c such that $6 + bx - 2x^2 = c - a(x-1)^2$.
12. Work out the values of p, q and r such that $5 - 12x - 2x^2 = p - q(x+r)^2$.
13. (i) Work out the values of a and b such that $x^2 - 8x + 20 = (x-a)^2 + b$.
    (ii) Hence make x the subject of $y = x^2 - 8x + 20$.
14. (i) Work out the values of p, q and r such that $3x^2 + 6x + 1 = p(x+q)^2 + r$.
    (ii) Hence make x the subject of $y = 3x^2 + 6x + 1$.

6 The minimum value of a quadratic expression

The minimum (or least) value of a quadratic expression can be found by completing the square.

The least value of an expression of the form $(x + a)^2 + b$ is b, and occurs when $x + a = 0$.

> Squared expressions cannot take a negative value, so $(x + a)^2 \geq 0$ and therefore $(x + a)^2 + b \geq b$.

Example 2.23

$x^2 + px + q$ can be rewritten as $\left(x + \frac{p}{2}\right)^2 - \left(\frac{p}{2}\right)^2 + q$.

(i) By completing the square find the least value of $x^2 + 6x - 10$.

(ii) What is the value of x when $x^2 + 6x - 10$ has its least value?

Solution

$x^2 + 6x - 10 = (x + 3)^2 - 3^2 - 10$

$= (x + 3)^2 - 19$

(i) The least value of $x^2 + 6x - 10$ is -19.

(ii) This occurs when $x + 3 = 0 \Rightarrow x = -3$.

If the coefficient of x^2 is not 1 then remove it as a factor of the first two terms before completing the square.

Example 2.24

(i) By completing the square find the least value of $2x^2 - 4x + 7$.

(ii) What is the value of x when $2x^2 - 4x + 7$ has its least value?

Solution

$2x^2 - 4x + 7 = 2\left[x^2 - 2x\right] + 7$ *Now complete the square on $x^2 - 2x$.*

$= 2\left[(x - 1)^2 - 1^2\right] + 7$

$= 2\left[(x - 1)^2 - 1\right] + 7$ *Now multiply throughout the outer bracket by 2, the factor which was removed at the beginning.*

$= 2(x - 1)^2 - 2 + 7$

$= 2(x - 1)^2 + 5$.

(i) The least value is 5.

(ii) It occurs when $x - 1 = 0 \Rightarrow x = 1$.

The minimum value of a quadratic expression

If the coefficient of x^2 is negative then the expression does not have a minimum value. Instead, it has a maximum (or greatest) value.

Example 2.25

(i) By completing the square find the least value of $-3x^2 + 8x - 2$

(ii) What is the value of x when $-3x^2 + 8x - 2$ has its least value?

Solution

$$-3x^2 + 7x - 2 = -3\left[x^2 - \frac{7}{3}x\right] - 2$$

$$= -3\left[\left(x - \frac{7}{6}\right)^2 - \left(\frac{7}{6}\right)^2\right] - 2$$

$$= -3\left[\left(x - \frac{7}{6}\right)^2 - \frac{49}{36}\right] - 2$$

$$= -3\left(x - \frac{7}{6}\right)^2 + \frac{49}{12} - \frac{24}{12}$$

$$= \frac{25}{12} - 3\left(x - \frac{7}{6}\right)^2$$

(i) The greatest value is $\frac{25}{12}$

(ii) It occurs when $x - \frac{7}{6} = 0 \implies x = \frac{7}{6}$

Exercise 2G

① Find the minimum values of the following expressions.

(i) $x^2 + 2x + 8$ (ii) $x^2 - 6x - 7$ (iii) $y^2 + 12y - 21$

(iv) $m^2 - 5m + 3$ (v) $n^2 + 7n + 1$ (vi) $x^2 - 11x$

② For each of these expressions, find the value of x for which the expression takes its least value.

(i) $x^2 - 4x + 9$ (ii) $x^2 + 10x - 5$ (iii) $x^2 + 9x + 6$

(iv) $x^2 - x + 7$ (v) $4 + 6x + x^2$ (vi) $x + x^2 - 5$

③ Find the minimum values of the following expressions.

(i) $2x^2 + 4x - 7$ (ii) $3x^2 - 6x + 1$ (iii) $4x^2 + 12x - 3$

(iv) $2x^2 - 5x + 4$ (v) $3x^2 - 4x + 2$ (vi) $5x^2 + 8x$

④ (i) Use the method of completing the square to find the least value of $x^2 + 18x + 100$.

(ii) What is the value of x when $x^2 + 18x + 100$ has its least value?

⑤ (i) Use the method of completing the square to find the greatest value of $-2x^2 - 8x + 5$.

(ii) What is the value of x when $-2x^2 - 8x + 5$ has its greatest value?

⑥ (i) Use the method of completing the square to find the greatest value of $-x^2 + 12x - 7$.

(ii) What is the value of x when $-x^2 + 12x - 7$ has its greatest value?

LEARNING OUTCOMES

Now you have finished this chapter, you should be able to

- factorise an algebraic expression using no more than two brackets
- rearrange a formula to make a different letter the subject
  - when the new subject occurs once
  - when the new subject appears more than once
- simplify algebraic fractions connected by any of the symbols $+, -, \times,$ or $\div$
- solve linear equations containing algebraic fractions
- write a quadratic expression in the form $a(x + b)^2 + c$.

KEY POINTS

1. When factorising a quadratic expression you need to write it as a product using brackets.
2. When changing the subject of an equation the new subject should be on its own on the left-hand side.
3. When simplifying an algebraic fraction involving addition or subtraction you need to find a common denominator.
4. When solving an equation involving fractions you start by multiplying through by the least common multiple of all the denominators to eliminate the fractions.
5. Quadratic expressions can be written in the form $a(x + b)^2 + c$.
6. The least (or greatest) value of $a(x + b)^2 + c$ is c, and occurs when $x = -b$.

3 Algebra III

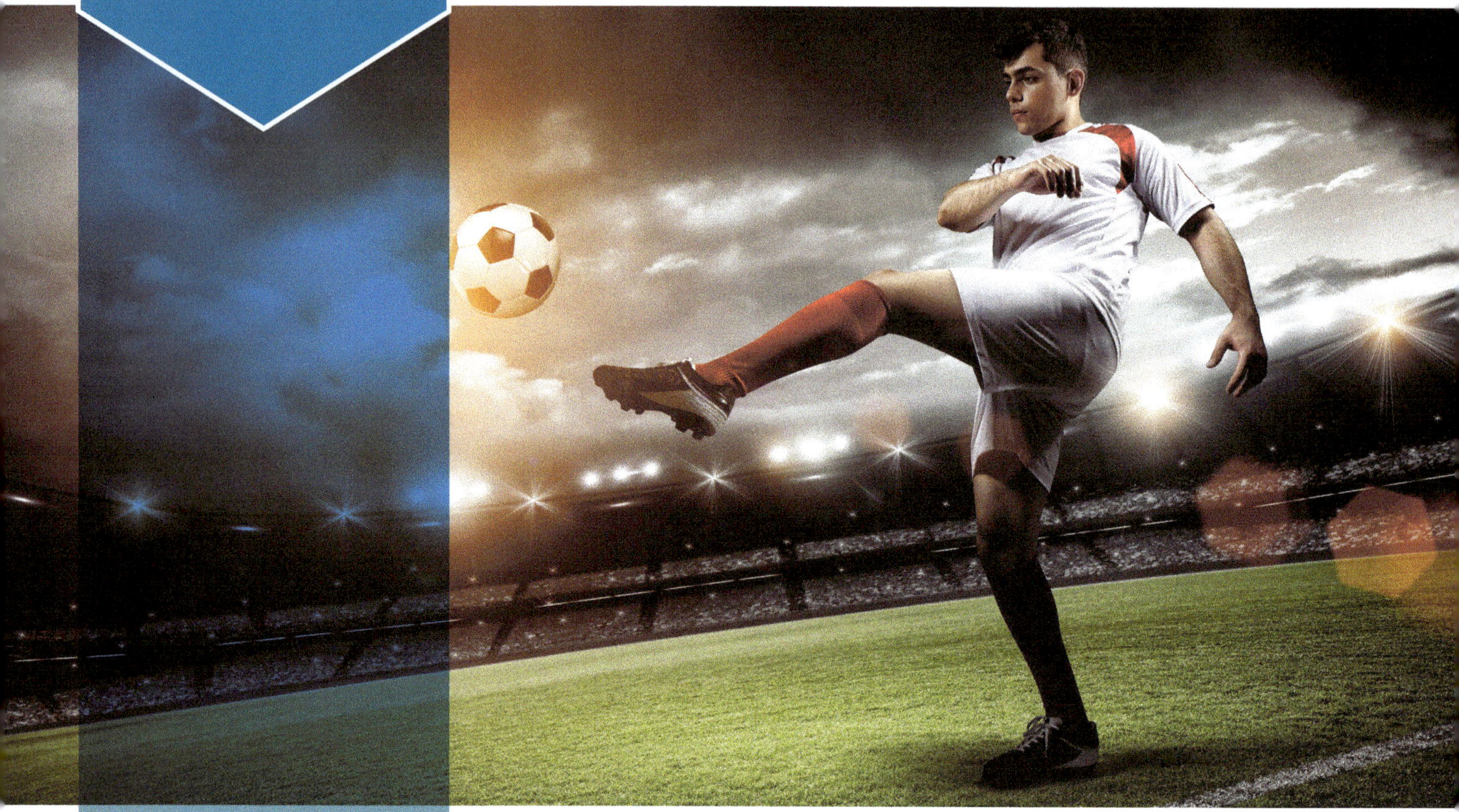

> Others have done it before me. I can, too.
>
> Corporal John Faunce
> (American soldier)

1 Function notation

Here is a flow chart.

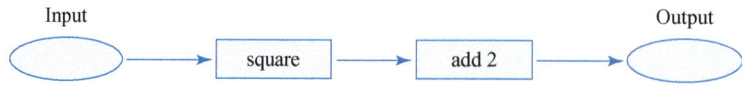

Figure 3.1

For an input of 5,	$5 \rightarrow 25 \rightarrow 27$	the output is 27.
For an input of −2,	$-2 \rightarrow 4 \rightarrow 6$	the output is 6.
For an input of x,	$x \rightarrow x^2 \rightarrow x^2 + 2$	the output is $x^2 + 2$.

This leads to the use of function notation $\quad f(x) = x^2 + 2$

For an input of 5, $\quad f(5) = 5^2 + 2$
$= 25 + 2$
$= 27.$

For an input of −2, $\quad f(-2) = (-2)^2 + 2$
$= 4 + 2$
$= 6.$

! A *function* must have a unique output for every input.

Consequently $f(x) = \pm x^2$ is not a function.

Discussion point

→ Which of the following are functions?
  (i) $f(x) = (\pm x)^2$
  (ii) $f(x) = (1 \pm x)^2$

Example 3.1

$f(x) = 10 - 4x$ and $g(x) = x^3$.

(i)   Evaluate $f(-1)$ and $g\left(\dfrac{1}{2}\right)$.

(ii) Write down an expression for $f(3x)$.

(iii) Solve $g(x) = -64$.

Solution

(i) $f(-1) = 10 - 4(-1)$
      $= 10 + 4$
      $= 14$

   $g\left(\dfrac{1}{2}\right) = \left(\dfrac{1}{2}\right)^3$
   $= \dfrac{1}{2} \times \dfrac{1}{2} \times \dfrac{1}{2}$
   $= \dfrac{1}{8}$

(ii) $f(3x) = 10 - 4(3x)$
      $= 10 - 12x$

(iii) $g(x) = -64$
      $x^3 = -64$
      $x = \sqrt[3]{-64}$
      $= -4$

Exercise 3A

① $f(x) = 2x - 1$ and $g(x) = x^2 + 2x$.
  Work out the value of
  (i) $f(-4)$
  (ii) $f(0.6)$
  (iii) $g(3)$
  (iv) $g(-1)$
  (v) $f(0)$
  (vi) $g(0)$.

② $f(x) = 3x^2$ and $g(x) = \dfrac{6}{x}$.
  Work out the value of
  (i) $f(2)$
  (ii) $f(-5)$
  (iii) $g(2)$
  (iv) $g(-1.5)$
  (v) $g\left(\dfrac{1}{2}\right)$
  (vi) $g\left(-\dfrac{2}{3}\right)$.

③ $f(x) = (2x - 1)^2$ and $g(x) = 2x + 1$.
  Work out the value of
  (i) $f(0)$
  (ii) $g(-2)$
  (iii) $f(0.5)$
  (iv) $f\left(-\dfrac{1}{4}\right)$
  (v) $g\left(-\dfrac{1}{2}\right)$
  (vi) $g(1.6)$.

④ $f(x) = 8 - 3x$ and $g(x) = 4(x + 3)$.
  Solve
  (i) $f(x) = 0$
  (ii) $g(x) = 20$
  (iii) $f(x) = g(x)$.

⑤ $h(x) = 3x - 2$
  Write down expressions, giving answers in the simplest form, for
  (i) $h(2x)$
  (ii) $h(x + 1)$
  (iii) $h(x^2)$.

⑥ $f(x) = (x-1)^2$

Write down expressions, giving answers in the simplest form, for

(i) $f(x^2)$ (ii) $[f(x)]^2$ (iii) $(f(x+1))^2$.

⑦ $f(x) = x^2 + 5x - 1$

Write down expressions, giving answers in the simplest form, for

(i) $f(3x)$ (ii) $f(x-2)$.

⑧ $g(x) = \dfrac{x+6}{2x}$

(i) Work out the value of $g(3)$.

(ii) Solve $g(x) = 3$.

(iii) Solve $g(2x) = 1$.

2 Graphs of functions

Drawing or plotting a graph

If asked to draw a graph you should use graph paper. The axes should be numbered. The graph should be drawn passing through the points which have either been given or calculated.

Here is a drawing of the graph of $y = 2x + 1$ for values of x from -2 to 4. In this case the coordinates of the points have been calculated.

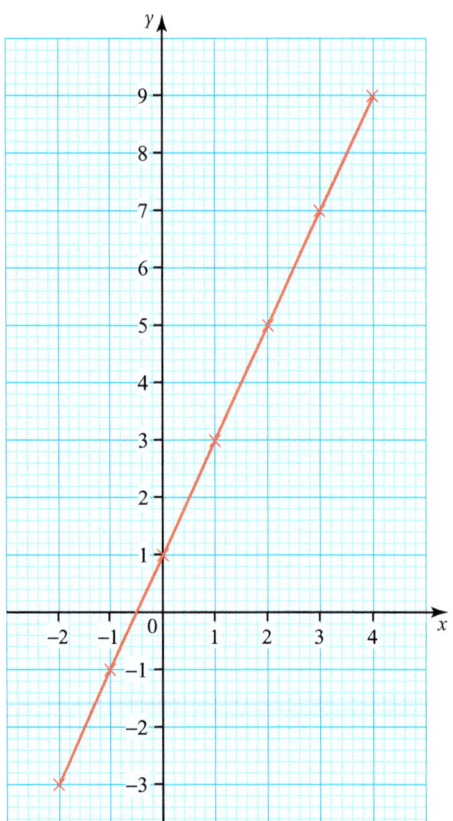

> **Note**
>
> In some cases you may meet the notation $f(x) = 2x + 1$ and then be told to draw the graph of $y = f(x)$. This is exactly the same instruction as 'Draw $y = 2x + 1$'.

Figure 3.2

Sketching a graph

If asked to sketch a graph you should *not* use graph paper. Axes should be drawn and only certain numbers need to be marked on the axes (e.g. points where the graph crosses the axes).

The correct shape of the graph should be shown and it should be in the correct position relative to the axes.

This means that the main features of the graph are shown although there is no requirement to plot points accurately.

Here is a sketch of the graph of $y = 2x + 1$

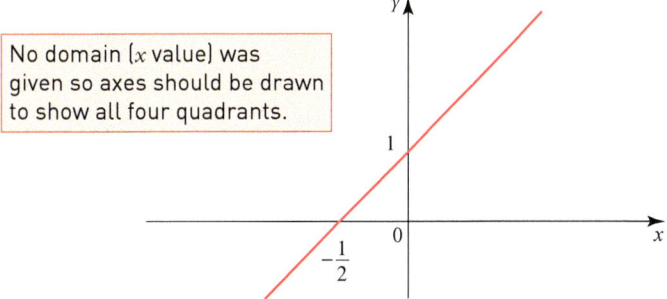

No domain (*x* value) was given so axes should be drawn to show all four quadrants.

Figure 3.3

3 Graphs of linear functions

The gradient of a line

In mathematics the word *linear* refers to a straight line. The slope of a line is measured by its *gradient* and the letter *m* is often used to represent this.

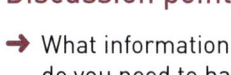

Discussion point

→ What information do you need to have in order to fix the position of a line?

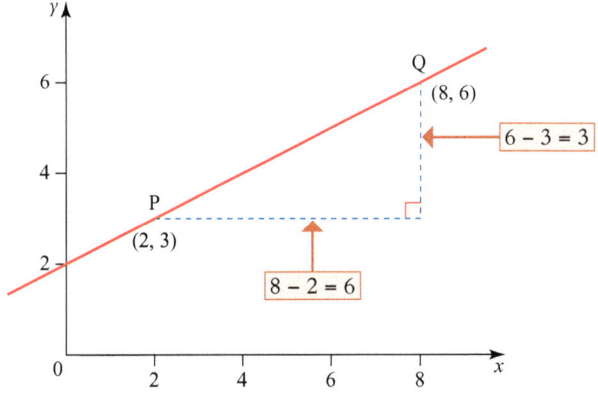

Figure 3.4

Gradient = $\dfrac{\text{change in } y\text{-coordinate from P to Q}}{\text{change in } x\text{-coordinate from P to Q}}$

In Figure 3.4, gradient = $\dfrac{6-3}{8-2} = \dfrac{3}{6} = \dfrac{1}{2}$.

Graphs of linear functions

ACTIVITY 3.1
On each line in Figure 3.5, take any two points and use them to calculate the gradient of the line.

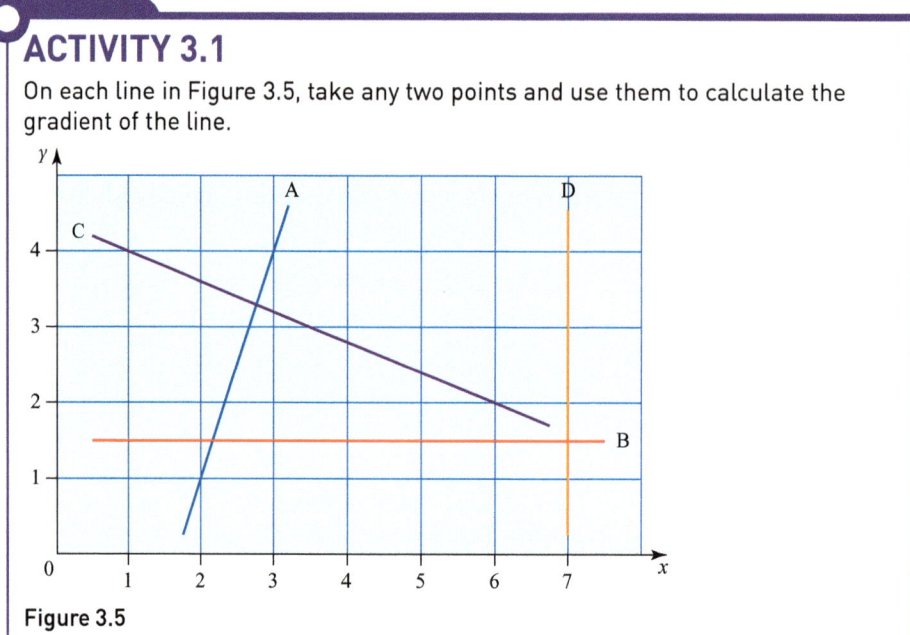

Figure 3.5

Discussion point
→ Does it matter which point you call (x_1, y_1) and which (x_2, y_2)?

You can generalise the previous activity to find the gradient m of the line joining (x_1, y_1) to (x_2, y_2).

$$m = \frac{y_2 - y_1}{x_2 - x_1}$$

You can easily tell by looking at a line if its gradient is positive, negative, zero or infinite.

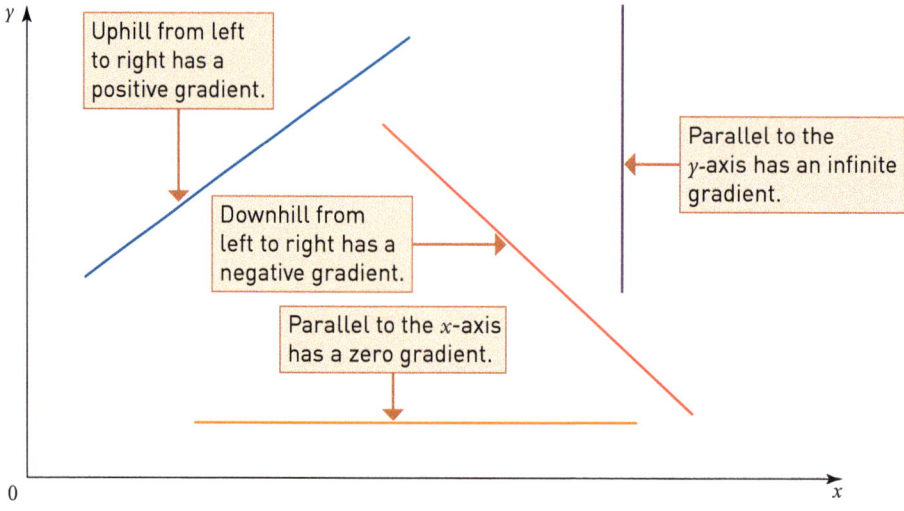

Figure 3.6

The equation of a straight line

Example 3.2

Work out the equation of the straight line with gradient 2 through the point with coordinates (0, 1).

Solution

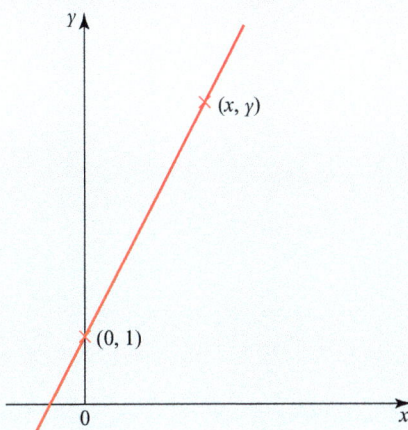

Figure 3.7

Take a general point (x, y) on the line, as shown in Figure 3.7, together with the point $(0, 1)$ that you are given. The gradient of the line joining $(0, 1)$ to (x, y) is given by

$$\text{gradient} = \frac{y - 1}{x - 0} = \frac{y - 1}{x}.$$

Since you are given that the gradient of the line is 2,

$$\frac{y - 1}{x} = 2 \quad \Rightarrow \quad y = 2x + 1$$

Since (x, y) is a general point on the line, this holds for any point on the line and is therefore the equation of the line.

This example can be generalised to give the result that the equation of the line with gradient m cutting the y-axis at the point $(0, c)$ is

$$\frac{y - c}{x - 0} = m$$

$$\Rightarrow \quad y = mx + c.$$

This is a well-known standard form for the equation of a straight line.

Graphs of linear functions

Drawing or sketching a line given its equation

There are several standard forms for the equation of a straight line (see Figure 3.8). When you need to draw or sketch a line, look at its equation and see if it fits one of these.

Equations of the form $x = a$

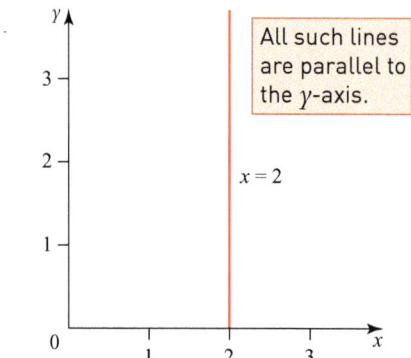

All such lines are parallel to the y-axis.

Equations of the form $y = b$

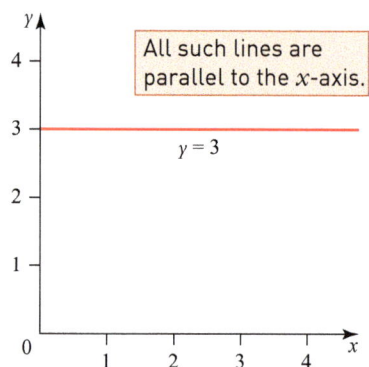

All such lines are parallel to the x-axis.

Equations of the form $y = mx$

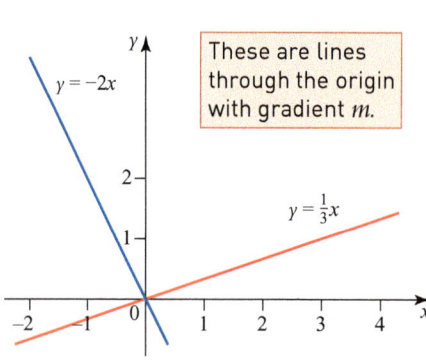

These are lines through the origin with gradient m.

Equations of the form $y = mx + c$

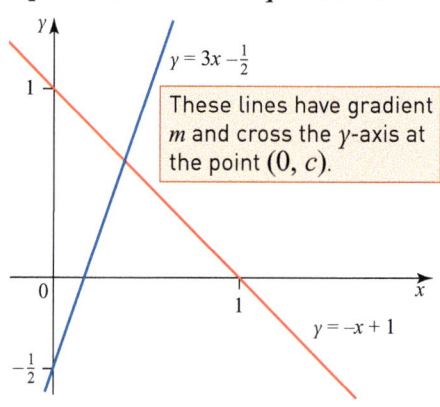

These lines have gradient m and cross the y-axis at the point $(0, c)$.

Equations of the form $px + qy + r = 0$

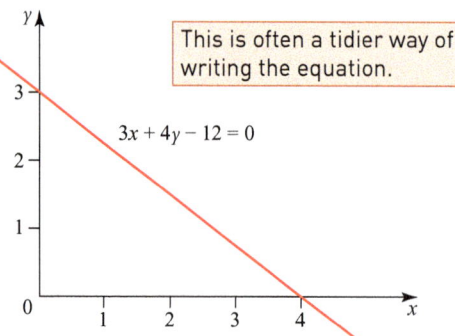

This is often a tidier way of writing the equation.

Figure 3.8

Graphs of equations in the form $px + qy + r = 0$ will usually be sketched by finding the coordinates of the points where the line crosses the x- and y-axes.

> **Discussion point**
>
> (i) Rearrange the equation $3x + 4y - 12 = 0$ into the form $\frac{x}{a} + \frac{y}{b} = 1$.
> (ii) What are the values of a and b?
> (iii) What do these numbers represent?

Example 3.3

Sketch the lines $y = -2$, $y = 3x - 2$ and $x + 3y - 9 = 0$ on the same axes.

Solution

The line $y = -2$ is parallel to the x-axis and passes through $(0, -2)$.

The line $y = 3x - 2$ has gradient 3 and passes through $(0, -2)$.

To sketch the line $x + 3y - 9 = 0$ find two points on it.

$x = 0 \quad \Rightarrow \quad 3y - 9 = 0 \quad \Rightarrow \quad y = 3 \quad (0, 3)$ is on the line.

$y = 0 \quad \Rightarrow \quad x - 9 = 0 \quad \Rightarrow \quad x = 9 \quad (9, 0)$ is on the line.

Figure 3.9 shows the three lines.

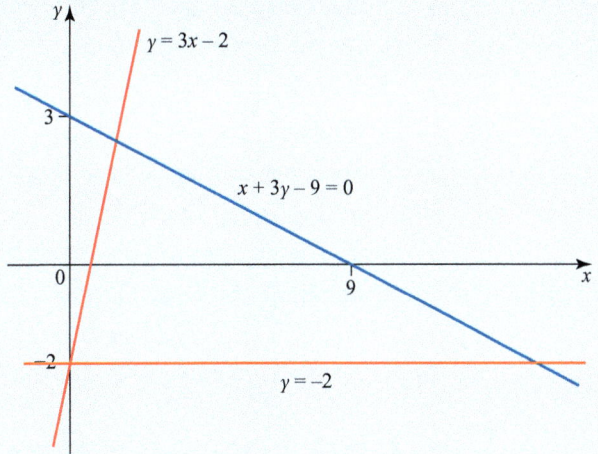

Figure 3.9

Exercise 3B

① For each of the following pairs of points, A and B, calculate the gradient of the line AB.

(i) A(4, 3)   B(8, 11)   (ii) A(3, 4)   B(0, 13)
(iii) A(5, 3) B(10, −8)   (iv) A(−6, −14) B(1, 7)
(v) A(6, 0)   B(8, 15)   (vi) A(−2, −4)   B(3, 9)
(vii) A(−3, −6) B(2, −7)   (viii) A(4, 7)   B(7, −4)

In the following questions mark the coordinates of all points of intersection with the axes.

② Sketch these lines.

(i) $x = 5$ (ii) $y = -3$ (iii) $x = 0$ (iv) $y = 0$

Finding the equation of a line

③ Sketch these lines.

(i) $y = 4x$ (ii) $y = -3x$ (iii) $y = 4 + x$ (iv) $y = -3 + x$

④ Sketch these lines.

(i) $y = 2x + 3$ (ii) $y = 2x - 3$ (iii) $y = 2 + 3x$ (iv) $y = 2 - 3x$

⑤ Sketch these lines.

(i) $y = \frac{1}{2}x - 1$ (ii) $y = \frac{1}{3}x + \frac{2}{3}$

(iii) $y = 2 - \frac{1}{2}x$ (iv) $y = 3 - \frac{2x}{3}$

⑥ Sketch these lines.

(i) $x + 2y = 5$ (ii) $3x - y = 4$

(iii) $2x + y = 0$ (iv) $x - 2y = 0$

⑦ Sketch these lines.

(i) $x + y - 1 = 0$ (ii) $2x + y - 4 = 0$

(iii) $x - 3y + 6 = 0$ (iv) $y - 3x + 9 = 0$

⑧ Sketch these lines.

(i) $\frac{x}{2} - \frac{y}{3} - 1 = 0$ (ii) $\frac{x}{3} - \frac{y}{2} - 1 = 0$

(iii) $\frac{3x}{2} - \frac{2y}{3} - 1 = 0$ (iv) $\frac{2x}{3} - \frac{3y}{2} - 1 = 0$

⑨ A printer quotes the cost £C of printing n business cards as $C = 60 + 0.06n$.

(i) Work out the cost of printing

(a) 500 cards

(b) 5000 cards

and the cost per card in each case.

(ii) The cost is made up of a fixed cost for setting up the printer and a cost per card printed. State the cost of each of these.

(iii) Sketch the graph of C against n.

If you have access to a graphic calculator, you can use it to check your results. Alternatively, check your answers with a free online graphing resource.

> **Prior knowledge**
>
> You should be confident when manipulating algebraic expressions, including algebraic fractions.

4 Finding the equation of a line

The simplest way of finding the equation of a straight line depends on what information you have been given.

Given the gradient, m, and the point of intersection $(0, c)$ with the y-axis

Take the general point (x, y) on the line, as shown in Figure 3.10.

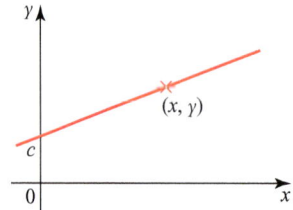

Figure 3.10

The gradient of the line joining $(0, c)$ to (x, y) is given by

$m = \dfrac{y - c}{x - 0}$

$\Rightarrow y = mx + c.$

Given the gradient, m, and the coordinates (x_1, y_1) of a point on the line

Take the general point (x, y) on the line, as shown in Figure 3.11.

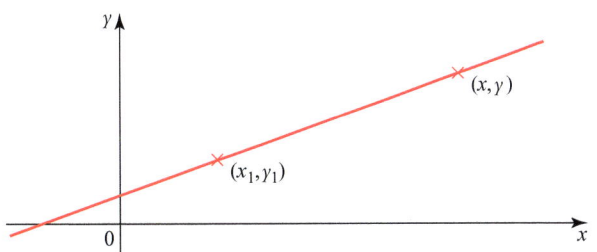

Figure 3.11

The gradient m of the line joining (x_1, y_1) to (x, y) is given by

$m = \dfrac{y - y_1}{x - x_1}$

$\Rightarrow y - y_1 = m(x - x_1).$

This is a standard result, and one you will find very useful.

Example 3.4

Work out the equation of the line with gradient 2 which passes through the point $(-1, 3)$.

Solution

Using $y - y_1 = m(x - x_1)$

$\Rightarrow y - 3 = 2(x - (-1))$

$\Rightarrow y - 3 = 2x + 2$

$\Rightarrow \quad y = 2x + 5$

In the formula

$y - y_1 = m(x - x_1)$

two positions of the point (x_1, y_1) lead to results you have met already:

- (x_1, y_1) is at $(0, 0) \Rightarrow y = mx$
- (x_1, y_1) is at $(0, c) \Rightarrow y = mx + c.$

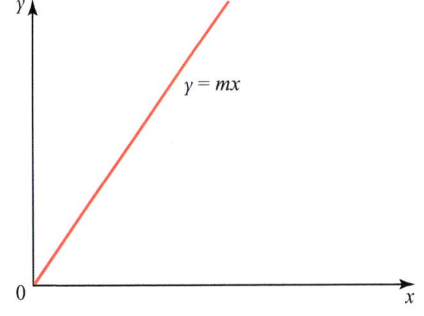 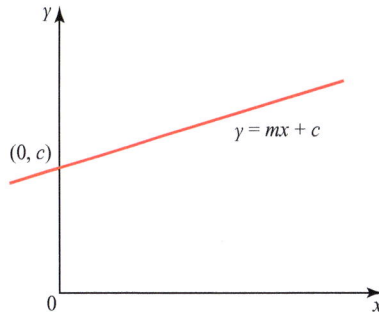

Figure 3.12

Finding the equation of a line

Given two points (x_1, y_1) and (x_2, y_2)

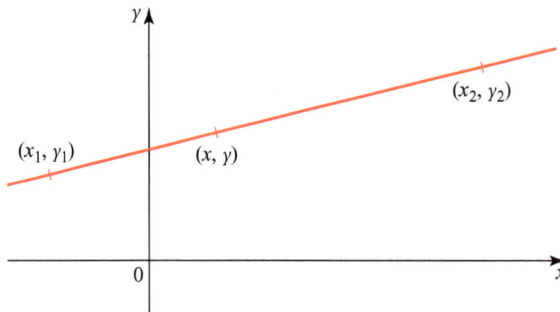

Figure 3.13

The two points are used to work out the gradient.

$$m = \frac{y_2 - y_1}{x_2 - x_1}$$

This value is then substituted in the equation

$$y - y_1 = m(x - x_1).$$

This gives

$$y - y_1 = \frac{y_2 - y_1}{x_2 - x_1}(x - x_1).$$

Rearranging this gives

$$\frac{y - y_1}{y_2 - y_1} = \frac{x - x_1}{x_2 - x_1} \quad \text{or}$$

$$\frac{y - y_1}{x - x_1} = \frac{y_2 - y_1}{x_2 - x_1}.$$

Example 3.5

Work out the equation of the line joining $(-1, 4)$ to $(2, -3)$.

Solution

Let (x_1, y_1) be $(-1, 4)$ and (x_2, y_2) be $(2, -3)$.

Substituting these values in $\dfrac{y - y_1}{y_2 - y_1} = \dfrac{x - x_1}{x_2 - x_1}$

gives $\qquad \dfrac{y - 4}{(-3) - 4} = \dfrac{x - (-1)}{2 - (-1)}$

$\Rightarrow \qquad \dfrac{y - 4}{(-7)} = \dfrac{x + 1}{3}$

$\Rightarrow \qquad 3(y - 4) = (-7)(x + 1)$

$\Rightarrow \qquad 7x + 3y - 5 = 0$

Applying the different techniques

Example 3.6 Work out the equations of the lines **(a)** to **(c)** in Figure 3.14.

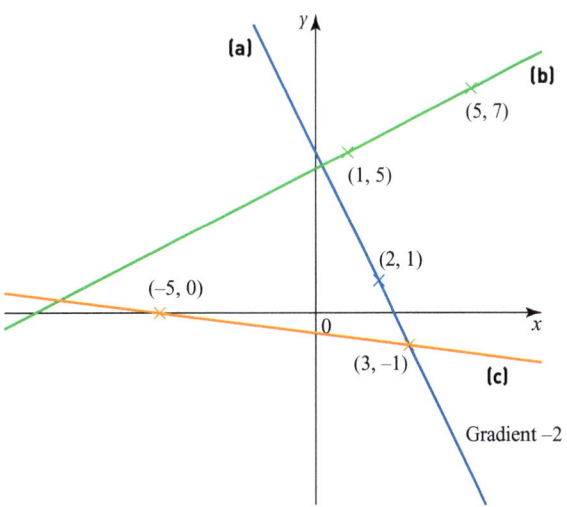

Figure 3.14

Solution

Line **(a)** has a gradient of −2 and passes through the point (2, 1).

Using $y - y_1 = m(x - x_1)$

$y - 1 = -2(x - 2)$

$\Rightarrow y - 1 = -2x + 4$

$\Rightarrow y = -2x + 5$

> This may also be written as $2x + y = 5$ or $2x + y - 5 = 0$.

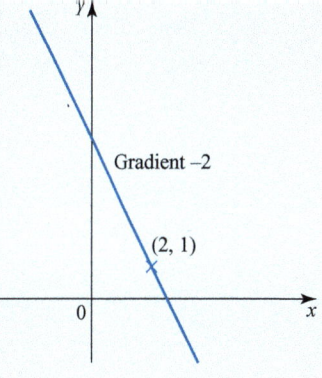

Figure 3.15

Line **(b)** passes through the points (1, 5) and (5, 7).

Using $m = \dfrac{y_2 - y_1}{x_2 - x_1}$ to find the gradient

$m = \dfrac{7 - 5}{5 - 1} = 0.5$

Using $y - y_1 = m(x - x_1)$ and the point (1, 5)

$y - 5 = 0.5(x - 1)$

$\Rightarrow y - 5 = 0.5x - 0.5$

$\Rightarrow y = 0.5x + 4.5$

> Avoiding a decimal in the answer, this could also be given as $2y = x + 9$ or $x - 2y + 9 = 0$.

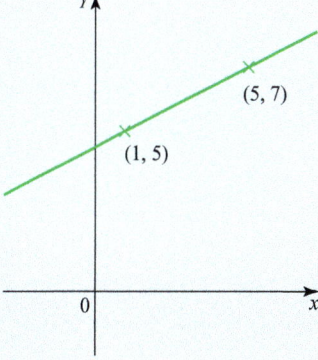

Figure 3.16

Finding the equation of a line

Line **(c)** passes through the points $(-5, 0)$ and $(3, -1)$.

Let (x_1, y_1) be $(-5, 0)$ and (x_2, y_2) be $(3, -1)$.

Substituting these values in
$$\frac{y - y_1}{y_2 - y_1} = \frac{x - x_1}{x_2 - x_1}$$

gives $\dfrac{y - 0}{-1 - 0} = \dfrac{x - (-5)}{3 - (-5)}$

$\Rightarrow \dfrac{y}{-1} = \dfrac{x + 5}{8}$

$\Rightarrow 8y = -x - 5$

$\Rightarrow x + 8y + 5 = 0$

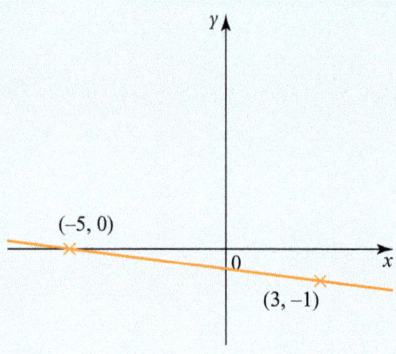

Figure 3.17

Exercise 3C

① Work out the equations of the lines **(i) – (v)** in this diagram.

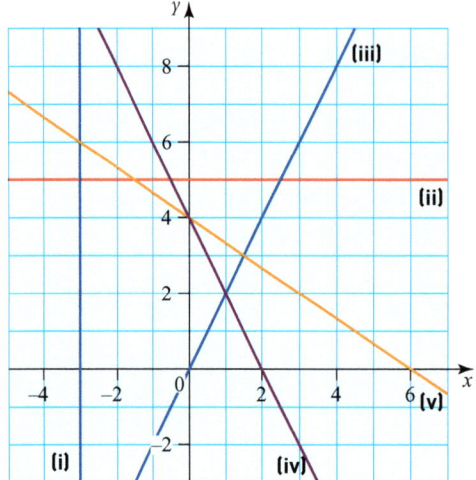

Figure 3.18

② Work out the equations of the lines **(i) – (v)** in this diagram.

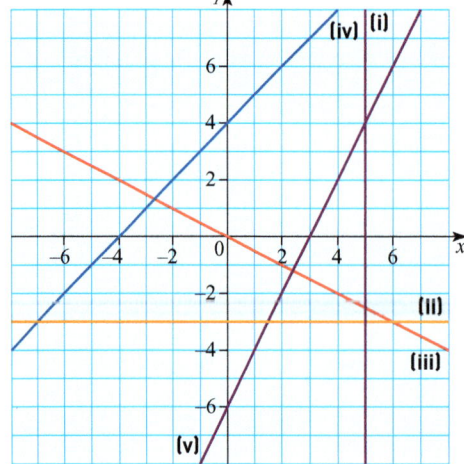

Figure 3.19

③ Work out the equations of these lines.
   (i) Gradient 3 and passing through (2, −1)
   (ii) Gradient 2 and passing through (0, 0)
   (iii) Gradient 3 and passing through (2, −7)
   (iv) Gradient 4 and passing through (4, 0)

④ Work out the equations of these lines.
   (i) Gradient $\frac{1}{3}$ and passing through (3, 1)
   (ii) Gradient $\frac{2}{5}$ and passing through (−4, −10)
   (iii) Gradient $-\frac{3}{2}$ and passing through (1, −2)
   (iv) Gradient $-\frac{1}{2}$ and passing through (0, 6)

⑤ Work out the equation of the line AB in each of these cases.
   (i) A(2, 0)   B(3, 1)
   (ii) A(3, −1) B(0, 4)
   (iii) A(2, −3) B(3, −2)

⑥ Work out the equation of the line AB in each of these cases.
   (i) A(−1, 3)   B(4, 0)
   (ii) A(3, 5)   B(10, 6)
   (iii) A(−1, −2) B(−4, −8)

⑦ A taxi journey costs £2 plus 80 pence per mile. Use £C to represent the total cost of the journey and m miles to represent the total distance travelled.
   (i) Write down an equation giving C in terms of m.
   (ii) How much would a journey of 4 miles cost?
   (iii) How far could I travel if I only had £10?

⑧ A junior school is ordering exercise books for their students and is working on the assumption that most students will only use 8 books during the year, but they want to order an additional 100. Let N represent the number of books to be ordered and let s be the number of students enrolled for the year.
   (i) Write down an equation giving N in terms of s.
   (ii) The exercise books cost £1.50 each. If there are 240 students that year, what would be the total cost of the books?
   (iii) The school budget for exercise books is only £3000. How could the order be amended?

5 Graphs of quadratic functions

Discussion point
→ What is the difference between a quadratic function and a quadratic equation?

ACTIVITY 3.2

Copy and complete the table of values and draw the graph of $y = x^2 - 5$ for values of x from −3 to 4.

x	−3	−2	−1	0	1	2	3	4
y	4			−5		−1	4	

Graphs of quadratic functions

ACTIVITY 3.3

Copy and complete the table of values and draw the graph of $y = 4x - x^2$ for values of x from -2 to 6.

x	-2	-1	0	1	2	3	4	5	6
y	-12	-5			4				-12

The shape of the graph of $y = ax^2 + bx + c$ is a parabola.

The sign of the coefficient of x^2 determines the direction of the curve.

$a > 0$

P is the lowest point on the graph in Figure 3.20.

P is the vertex.

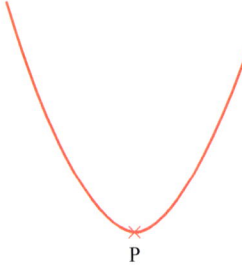

Figure 3.20

$a < 0$

Q is the highest point on the graph in Figure 3.21.

Q is the vertex.

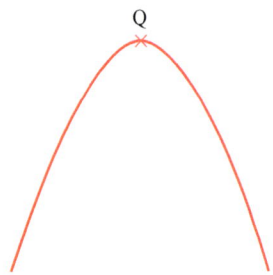

Figure 3.21

Symmetry

Quadratic graphs have a line of symmetry when drawn using an appropriate domain.

Here is the graph of $y = x^2 - 2x - 3$ for domain $-2 \leq x \leq 4$.

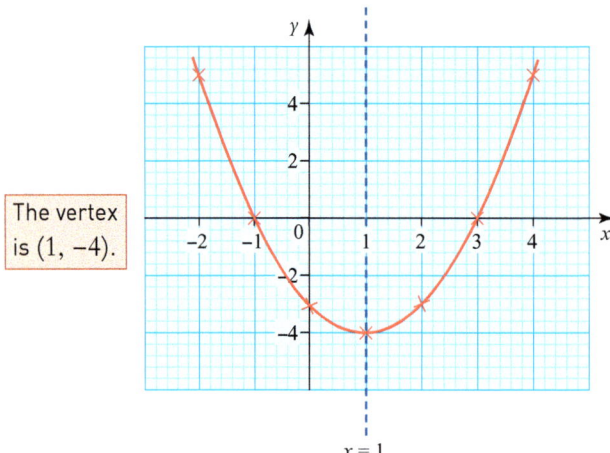

The vertex is $(1, -4)$.

Figure 3.22

The line of symmetry has equation $x = 1$ and passes through the vertex.

Here is a sketch of the graph of $y = 9 - x^2$ for domain $-4 \leq x \leq 4$.

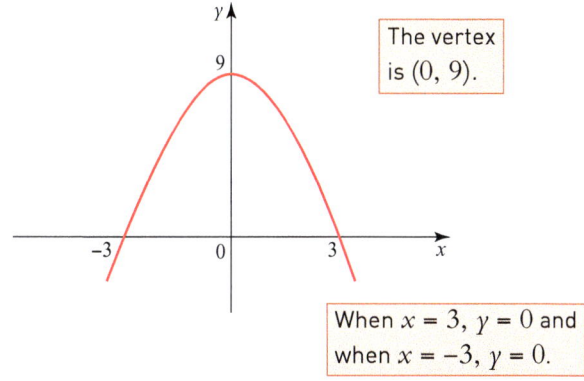

The vertex is (0, 9).

When $x = 3$, $y = 0$ and when $x = -3$, $y = 0$.

Figure 3.23

The line of symmetry is the y-axis and passes through the vertex.

Prior knowledge

You are expected to be familiar with the technique of completing the square, used to find the vertex and the line of symmetry of a quadratic graph, from GCSE. This is used in the following example.

This technique is covered in more detail in the next chapter.

Example 3.7

For the graph of $y = x^2 + 6x + 11$, state

(i) the vertex

(ii) the equation of the line of symmetry

(iii) the coordinates of the point where the graph intersects the y-axis.

Sketch the graph.

Solution

Write the quadratic expression in the form $(x + a)^2 + b$.

$x^2 + 6x + 11 \equiv (x + a)^2 + b$

$\equiv x^2 + 2ax + a^2 + b$

Equate coefficients of x means making equal the number of x on each side of the identity.

Equate coefficients of x $6 = 2a$

$3 = a$

Equate constants $11 = a^2 + b$

$11 = 9 + b$

$2 = b$

$(x + 3)^2$ is always positive or zero.

The least value of $(x + 3)^2 + 2$ is 2 and this occurs when $x = -3$.

$y = x^2 + 6x + 11$
$= (x + 3)^2 + 2$

(i) The vertex is $(-3, 2)$.

(ii) The equation of the line of symmetry is $x = -3$.

(iii) When $x = 0$, $y = 11$, so coordinates of y-axis intercept are $(0, 11)$.

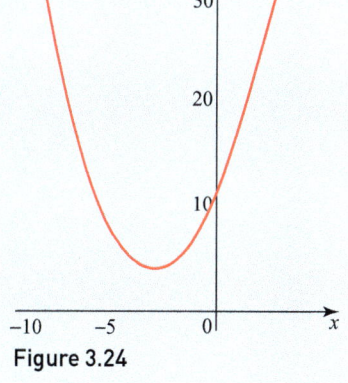

Figure 3.24

Graphs of quadratic functions

Exercise 3D

① Choose an equation from the list below to fit each of the quadratic curves **(i)** and **(ii)** in Figure 3.25.

$y = x^2 - 2x + 4$
$y = 5 - x^2$
$y = 3x - x^2$
$y = x^2 - 2x - 3$

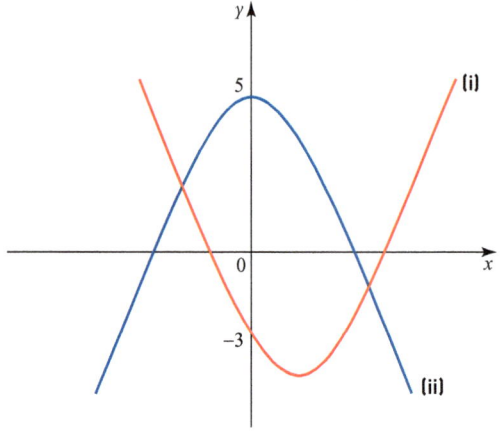

Figure 3.25

② Choose an equation from the list below to fit each of the quadratic curves **(i)** and **(ii)** in Figure 3.26.

$y = x^2 + 3x + 4$
$y = 4 - 7x - 2x^2$
$y = x^2 + 2x$
$y = 4x - x^2$

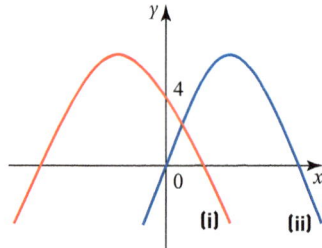

Figure 3.26

③ (i) For the graph of $y = x^2 + 2x + 3$, work out
  (a) the vertex
  (b) the equation of the line of symmetry
  (c) the coordinates of the point where the graph intersects the y-axis.
  (ii) Sketch the graph.

④ (i) For the graph of $y = x^2 - 4x + 5$, work out
  (a) the vertex
  (b) the equation of the line of symmetry
  (c) the coordinates of the point where the graph intersects the y-axis.
  (ii) Sketch the graph.

⑤ (i) For the graph of $y = x^2 - 6x + 7$, work out
  (a) the vertex
  (b) the equation of the line of symmetry
  (c) the coordinates of the point where the graph intersects the y-axis.
  (ii) Sketch the graph.

⑥ (i) For the graph of $y = x^2 - 3x - 4$, work out
  (a) the vertex
  (b) the equation of the line of symmetry
  (c) the coordinates of the point where the graph intersects the y-axis.

(ii) Use trial and improvement to work out the coordinates of the points where the curve intersects the x-axis.

(iii) Sketch the graph.

⑦ (i) Show that $x^2 - 6x + 10 = (x - 3)^2 + 1$.

(ii) Write down the vertex of the graph of $y = x^2 - 6x + 10$.

(iii) Write down the equation of the line of symmetry of $y = x^2 - 6x + 10$.

(iv) Write down the coordinates of the points where the graph of $y = x^2 - 6x + 10$ intersects the coordinate axes.

(v) Sketch the graph of $y = x^2 - 6x + 10$.

⑧ (i) Show that $6 - 4x - x^2 = 10 - (x + 2)^2$.

(ii) Write down the vertex of the graph of $y = 6 - 4x - x^2$.

(iii) Write down the equation of the line of symmetry of $y = 6 - 4x - x^2$.

(iv) Write down the coordinates of the points where the graph of $y = 6 - 4x - x^2$ intersects the y-axis.

(v) Sketch the graph of $y = 6 - 4x - x^2$.

⑨ A quadratic curve has its vertex at $(3, -7)$ and passes through $(0, 2)$.

Work out the equation of the curve.

⑩ A car hire company is negotiating the price of n identical new cars with the manufacturer and offers a total price £P, where $P = 20\,000n - 200n^2$.

(i) What is the likely cost of one car?

(ii) Work out the cost of an order for

(a) 5 cars (b) 20 cars (c) 50 cars.

(iii) Work out the average cost per car when the number of cars ordered is

(a) 5 cars (b) 20 cars (c) 50 cars.

(iv) Suggest why the manufacturer may not want to sell 50 cars at once.

LEARNING OUTCOMES

Now you have finished this chapter, you should be able to
- draw the graph of a function on graph paper
- sketch a graph – not using graph paper
- find the equation of a line given either the coordinates of two points on the line or the gradient of the line and the coordinates of one point
- recognise the shape of a quadratic graph from its equation, determining whether it has a maximum or minimum turning point.

KEY POINTS

1. A function maps an input, x, to an output, $f(x)$.
2. When asked to draw a graph, use graph paper.
3. When asked to sketch a graph, do not use graph paper.
4. The gradient of the straight line joining the points (x_1, y_1) and (x_2, y_2) is given by $\dfrac{y_2 - y_1}{x_2 - x_1}$.

Key points

5 The equation of a straight line may take any of these forms.
- Line parallel to the y-axis: $x = a$
- Line parallel to the x-axis: $y = b$
- Line through the origin with gradient m: $y = mx$
- Line through $(0, c)$ with gradient m: $y = mx + c$
- Line through (x_1, y_1) with gradient m: $y - y_1 = m(x - x_1)$
- Line through (x_1, y_1) and (x_2, y_2):
$$\frac{y - y_1}{y_2 - y_1} = \frac{x - x_1}{x_2 - x_1} \quad \text{or} \quad \frac{y - y_1}{x - x_1} = \frac{y_2 - y_1}{x_2 - x_1}$$

6 The shape of a quadratic graph is a parabola. It may be either $\cup$ shaped (when the coefficient of the squared term is positive) or $\cap$ shaped (when it is negative).

4 Algebra IV

'Obvious' is the most dangerous word in mathematics.

E. T. Bell

1 Quadratic equations

Solving a quadratic equation by factorising

When solving an equation by factorisation it is essential that all non-zero terms are moved to one side of the equation, leaving zero on the other side. This is due to a unique property of zero: when the product of two (or more) numbers or expressions is zero, then at least one of the numbers or expressions must be zero. No other number has such a property.

Example 4.1

Solve $x^2 = 4x + 21$.

Solution

$$x^2 = 4x + 21$$
$$\Rightarrow x^2 - 4x - 21 = 0$$
$$\Rightarrow (x + 3)(x - 7) = 0$$
$$\Rightarrow x + 3 = 0 \text{ or } x - 7 = 0$$
$$\Rightarrow x = -3 \text{ or } x = 7$$

Quadratic equations

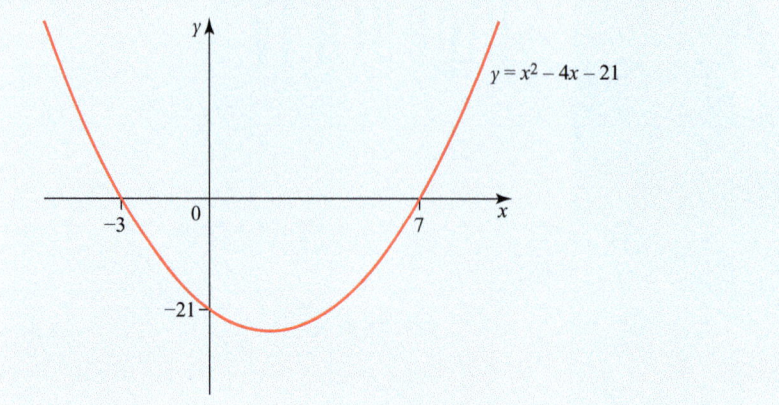

Figure 4.1

> ⚠ Before solving a quadratic equation by factorisation, ensure that all non-zero terms are on one side of the equation only.

Example 4.2

Solve $8x^2 + 10x = 3$.

Solution

First, move the 3 to the left-hand side of the equation, leaving zero on the other side.

$$8x^2 + 10x = 3$$
$$\Rightarrow 8x^2 + 10x - 3 = 0$$
$$\Rightarrow 8x^2 + 12x - 2x - 3 = 0$$
$$\Rightarrow 4x(2x + 3) - 1(2x + 3) = 0$$
$$\Rightarrow (2x + 3)(4x - 1) = 0$$
$$\Rightarrow 2x + 3 = 0 \quad \text{or} \quad 4x - 1 = 0$$
$$\Rightarrow x = -\frac{3}{2} \quad \text{or} \quad x = \frac{1}{4}$$

> Find two numbers whose sum is 10 and product is −24 (10 is the coefficient of x, and −24 is the product of the constant term and the coefficient of x^2).
> The numbers required are 12 and −2.

Sometimes a quadratic equation will not factorise.

In this case, you must complete the square or use the quadratic formula.

Solving a quadratic equation by completing the square

Example 4.3

Solve $x^2 - 8x + 3 = 0$.

Solution

> Consider the coefficient of x (−8).
> Halve it (−4).
> Then square the half (16).
> We then add this to both sides of the equation. This process is called 'completing the square'.

Subtract the constant term from both sides of the equation.

$$\Rightarrow x^2 - 8x = -3$$

Add 16 to both sides of the equation $\Rightarrow x^2 - 8x + 16 = -3 + 16$

Factorise the left-hand side $\Rightarrow (x - 4)^2 = 13$

Square root both sides $\Rightarrow x - 4 = \pm\sqrt{13}$

Add 4 to both sides $\Rightarrow x = 4 + \sqrt{13} \quad \text{or} \quad x = 4 - \sqrt{13}$

$$\Rightarrow x = 7.6055... \quad \text{or} \quad x = 0.3944...$$

> If the square has been completed correctly, the left-hand side will always factorise to the form $(x \pm p)^2$.

If the coefficient of the squared term is not 1, then first divide the equation by the coefficient.

Example 4.4 Solve $2x^2 + 3x - 7 = 0$.

Solution

$$2x^2 + 3x - 7 = 0$$
$$\Rightarrow x^2 + \frac{3}{2}x - \frac{7}{2} = 0$$
$$\Rightarrow x^2 + \frac{3}{2}x = \frac{7}{2}$$
$$\Rightarrow \left(x + \frac{3}{4}\right)^2 - \left(\frac{3}{4}\right)^2 = \frac{7}{2}$$
$$\Rightarrow \left(x + \frac{3}{4}\right)^2 - \frac{9}{16} = \frac{7}{2}$$
$$\Rightarrow \left(x + \frac{3}{4}\right)^2 = \frac{56}{16} + \frac{9}{16}$$
$$\Rightarrow \left(x + \frac{3}{4}\right)^2 = \frac{65}{16}$$
$$\Rightarrow x + \frac{3}{4} = \pm\sqrt{\frac{65}{16}}$$
$$\Rightarrow x = -\frac{3}{4} \pm \frac{\sqrt{65}}{4}$$

The quadratic formula

$$ax^2 + bx + c = 0$$
$$\Rightarrow x^2 + \frac{b}{a}x + \frac{c}{a} = 0$$
$$\Rightarrow x^2 + \frac{b}{a}x = -\frac{c}{a}$$
$$\Rightarrow \left(x + \frac{b}{2a}\right)^2 - \left(\frac{b}{2a}\right)^2 = -\frac{c}{a}$$
$$\Rightarrow \left(x + \frac{b}{2a}\right)^2 - \frac{b^2}{4a^2} = -\frac{c}{a}$$
$$\Rightarrow \left(x + \frac{b}{2a}\right)^2 = -\frac{4ac}{4a^2} + \frac{b^2}{4a^2}$$
$$\Rightarrow \left(x + \frac{b}{2a}\right)^2 = \frac{b^2 - 4ac}{4a^2}$$
$$\Rightarrow x + \frac{b}{2a} = \pm\sqrt{\frac{b^2 - 4ac}{4a^2}}$$
$$\Rightarrow x = -\frac{b}{2a} \pm \frac{\sqrt{b^2 - 4ac}}{2a}$$
$$\Rightarrow x = \frac{-b \pm \sqrt{b^2 - 4ac}}{2a}$$

Discussion points
- If $b^2 - 4ac$ is zero, how are the answers affected?
- If $b^2 - 4ac$ is negative, how are the answers affected?

The result $x = \dfrac{-b \pm \sqrt{b^2 - 4ac}}{2a}$ is known as the *quadratic formula*. It can be used to solve any quadratic equation.

The $\pm$ sign indicates that there are two possible roots. One root is found by using the $+$ sign, and the other by using the $-$ sign.

Quadratic equations

Figure 4.2 shows a parabola – the shape of a quadratic curve. The dotted line is the line of symmetry.

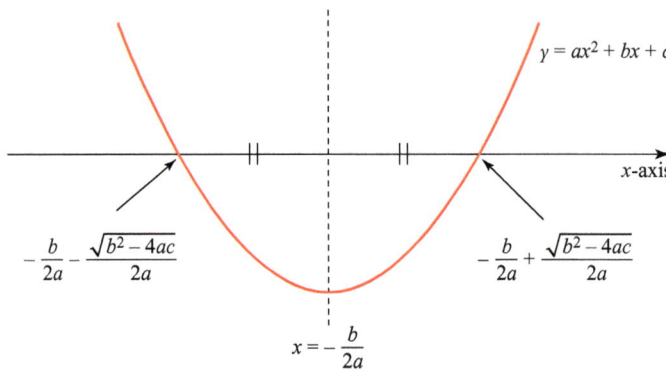

Figure 4.2

Solving a quadratic equation using the quadratic formula

Example 4.5

Use the quadratic formula to solve $3x^2 - 4x - 2 = 0$.

Solution

Comparing

$$3x^2 - 4x - 2 = 0$$

with

$$ax^2 + bx + c = 0$$

gives

$a = 3, \ b = -4, \ c = -2$.

Using these values in the formula gives

$$x = \frac{-b \pm \sqrt{b^2 - 4ac}}{2a}$$

$$x = \frac{-(-4) \pm \sqrt{(-4)^2 - 4 \times 3 \times -2}}{2 \times 3}$$

$$x = \frac{4 \pm \sqrt{16 + 24}}{6}$$

$$x = \frac{4 \pm \sqrt{40}}{6}.$$

In a non-calculator paper, the answer could be simplified and left in an exact form as

$$x = \frac{2 \pm \sqrt{10}}{3}.$$

Alternatively, in a calculator paper, approximate roots could be calculated as

$x = 1.72, \quad x = -0.39 \quad$ (rounded to 2 d.p.).

Example 4.6

The length of a carpet is 1 m greater than its width. Its area is 9 m².

Work out the dimensions of the carpet to the nearest centimetre.

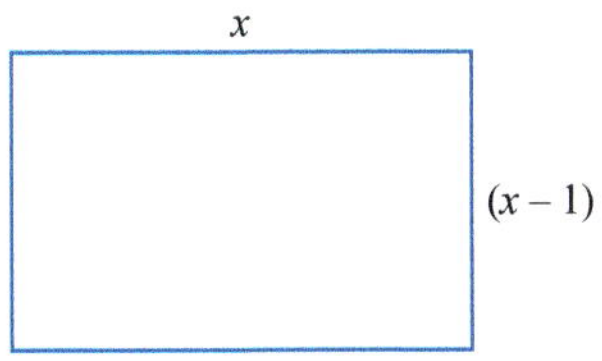

Figure 4.3

Solution

Let the length be x metres, so the width is $(x - 1)$ metres.

length × width = area

so $\quad x(x - 1) = 9$

$\Rightarrow \quad x^2 - x = 9$

$\Rightarrow \quad x^2 - x - 9 = 0 \quad$ (collecting all terms on the left-hand side)

Substituting $a = 1, b = -1, c = -9$ into the formula

$$x = \frac{-b \pm \sqrt{b^2 - 4ac}}{2a}$$

gives $\quad x = \dfrac{-(-1) \pm \sqrt{(-1)^2 - 4 \times 1 \times (-9)}}{2 \times 1}$

$\Rightarrow \quad x = \dfrac{1 \pm \sqrt{37}}{2}$

$\Rightarrow \quad x = 3.541...$ or $x = -2.541...$

Clearly a negative answer is not feasible, so the dimensions are

length = 3.54 m

width = 2.54 m   (to the nearest cm).

Example 4.7

Solve

$$\frac{5}{a + 1} - \frac{2a}{a^2 - 1} = \frac{1}{2}.$$

Solution

First factorise $(a^2 - 1)$ as $(a + 1)(a - 1)$.

$$\frac{5}{a + 1} - \frac{2a}{(a + 1)(a - 1)} = \frac{1}{2}$$

Multiply each term by $2(a + 1)(a - 1)$.

$\Rightarrow \quad 2(a+1)(a-1) \times \dfrac{5}{(a+1)} - 2(a+1)(a-1) \times \dfrac{2a}{(a+1)(a-1)}$

$\qquad\qquad\qquad\qquad = 2(a+1)(a-1) \times \dfrac{1}{2}$

$\Rightarrow \quad 10(a - 1) - 4a = (a + 1)(a - 1)$

$\Rightarrow \quad 10a - 10 - 4a = a^2 - 1$

$\Rightarrow \quad 0 = a^2 - 6a + 9$

$\Rightarrow \quad 0 = (a - 3)(a - 3)$

$\Rightarrow \quad a = 3$ (repeated root)

Quadratic equations

Formulation and manipulation of quadratic equations

When solving problems using algebra, let an unknown value be x.

Write all other unknown values in terms of x.

Form an equation using the information given in the question.

If the equation is quadratic, then rearrange it into the form $ax^2 + bx + c = 0$.

Such equations will often have two solutions. One of the solutions can usually be rejected because it is not a sensible answer. For example, negative or non-integer solutions are often nonsensical in some situations.

Example 4.8

When Dylan doubles his age (x) and then adds 3, the result is the same as Bethan's age.

The product of their ages is 90.

(i) Write down a quadratic equation in terms of x.

(ii) Hence find Dylan's age.

Solution

(i) $x(2x + 3) = 90$ ← Bethan's age is $2x + 3$.

$2x^2 + 3x = 90$

$2x^2 + 3x - 90 = 0$

(ii) Find two numbers with a sum of 3 and a product of -180. ← $2 \times (-90) = -180$

$2x^2 + 15x - 12x - 90 = 0$ ← $15 + (-12) = 3$ and $15 \times (-12) = -180$

$x(2x + 15) - 6(2x + 15) = 0$

$(2x + 15)(x - 6) = 0$

$x = -7.5$ or $x = 6$

Dylan's age is 6. ← A person's age must be positive, so -7.5 can be rejected.

Exercise 4A

1. Solve the following equations by factorising.

    (i) $x^2 - 8x + 12 = 0$
    (ii) $m^2 - 4m + 4 = 0$
    (iii) $p^2 - 2p - 15 = 0$
    (iv) $a^2 + 11a + 18 = 0$
    (v) $2x^2 + 5x + 2 = 0$
    (vi) $4x^2 + 3x - 7 = 0$
    (vii) $15t^2 + 2t - 1 = 0$
    (viii) $24r^2 + 19r + 2 = 0$
    (ix) $3x^2 + 8x = 3$
    (x) $3p^2 = 14p - 8$

② Solve the following equations
  (a) by completing the square
  (b) by using the quadratic formula.
  Give your answers correct to 2 decimal places.
  (i) $x^2 - 2x - 10 = 0$
  (ii) $x^2 + 3x - 6 = 0$
  (iii) $x^2 + x - 8 = 0$
  (iv) $2x^2 + x - 8 = 0$
  (v) $2x^2 + 2x - 9 = 0$
  (vi) $x^2 + x = 10$
  (vii) $x^2 = 4x + 1$
  (viii) $2x^2 - 8x + 5 = 0$

③ Solve the following equations by using the quadratic formula.
  Give your answers correct to 2 decimal places.
  (i) $3x^2 + 5x = -1$
  (ii) $4x^2 = -9x - 3$
  (iii) $2x^2 + 11x = 4$
  (iv) $4x^2 + 4 = 9x$
  (v) $5x^2 + 1 = 10x$
  (vi) $-9 - 11x = 3x^2$

④ The sides of a rectangle are x cm and $(x + 1)$ cm.
  The area of the rectangle is $110\,\text{cm}^2$.
  (i) Form a quadratic equation in terms of x.
  (ii) Solve the quadratic equation by factorising.
  (iii) Calculate the perimeter of the rectangle.

⑤ The sides of a right angled triangle, in centimetres, are x, $2x - 2$, and $x + 2$, where $x + 2$ is the hypotenuse.
  Use Pythagoras' theorem to work out their lengths.

SA ⑥ A rectangular lawn measures 8 m by 10 m and is surrounded by a path of uniform width x m. The total area of the path is $63\,\text{m}^2$.
  Work out the value of x.

IR ⑦ The difference between two positive numbers is 2 and the difference between their squares is 40.
  Work out the two numbers.

⑧ The formula $h = 15t - 5t^2$ gives the height h metres of a ball, t seconds after it is thrown up into the air.
  (i) Work out the times when the height is 10 m.
  (ii) After how long does the ball hit the ground?

⑨ The area of this triangle is $68\,\text{cm}^2$.

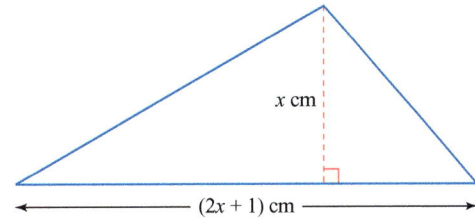

Figure 4.4

  (i) Show that x satisfies the equation $2x^2 + x - 136 = 0$.
  (ii) Solve the equation to work out the length of the base of the triangle.

Quadratic equations

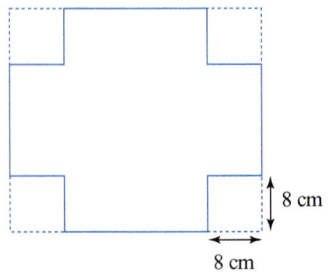

Figure 4.5

10. Boxes are made by cutting squares of side 8 cm from the corners of rectangular sheets of cardboard and then folding the remaining card. The sheets of cardboard are 6 cm longer than they are wide.
    (i) For a sheet of cardboard with width x cm, write expressions, in terms of x, for
        (a) the length of the sheet
        (b) the length of the finished box
        (c) the width of the finished box.
    (ii) Show that the volume of the box is $8x^2 - 208x + 1280$ cm³.
    (iii) Work out the dimensions of the sheet of cardboard needed to make a box with a volume of 1728 cm³.

11. Solve the following equations.
    (i) $\dfrac{1}{x} = 3 - \dfrac{2}{x+1}$
    (ii) $\dfrac{2}{3x-1} + \dfrac{1}{x+8} = \dfrac{1}{2}$
    (iii) $\dfrac{2}{a} - \dfrac{5}{2a-1} = 0$
    (iv) $\dfrac{6}{p-2} + \dfrac{6}{p+1} = 1$

12. Solve the following equations.
    (i) $\dfrac{1}{p} + p + 1 = \dfrac{13}{3}$
    (ii) $1 + \dfrac{1}{x-1} = \dfrac{2x}{x+1}$
    (iii) $\dfrac{6r}{r+1} - \dfrac{5}{r+3} = 3$
    (iv) $\dfrac{2}{x-3} + \dfrac{1}{x-1} = \dfrac{5}{4}$
    (v) $\dfrac{2}{x-1} - \dfrac{3}{x+1} = \dfrac{1}{4}$
    (vi) $\dfrac{5}{x+2} = 3 - \dfrac{4}{x-1}$
    (vii) $\dfrac{1}{2x+1} = \dfrac{4}{3} + \dfrac{1}{x-2}$
    (viii) $\dfrac{7}{3x-2} - \dfrac{1}{x-1} = \dfrac{1}{2}$
    (ix) $\dfrac{2}{2x-1} = 1 + \dfrac{3}{2x+1}$

13. Solve the following equations.
    (i) $\dfrac{a+4}{2a-3} = \dfrac{3(a+7)}{4(a+2)}$
    (ii) $\dfrac{4x-13}{2x+1} = \dfrac{5x-23}{x+5}$
    (iii) $\dfrac{2x+7}{x+7} = \dfrac{5x+13}{3-x}$

SA 14. A formula used in physics is $\dfrac{1}{f} = \dfrac{1}{u} + \dfrac{1}{v}$ where f is the focal length of a mirror, u is the distance of the object from the mirror, and v is the distance of the image from the mirror.

    For a mirror with focal length 20 cm, work out the distance of the object from the mirror when the image is twice as far away from the mirror as the object is.

IR 15. Anna has used the quadratic formula to solve a quadratic equation.

    She correctly calculated the answers as $x = \dfrac{4 \pm \sqrt{124}}{6}$.

    Write down an equation that Anna might have solved.

16. One edge of a square-based cuboid is 5 cm longer than another.
    The surface area of the cuboid is 114 cm².
    Find the two possible volumes of the cuboid.

2 Simultaneous equations in two unknowns

The equations you have met so far have involved only one unknown.

For example, $2x + 2 = x - 5$ or $a^2 - 3a + 2 = 0$.

Figure 4.6 shows the line $x + y = 4$.

> **Discussion point**
> → When an equation involves two unknowns, for example $x + y = 4$, how many possible pairs of values are there for x and y?

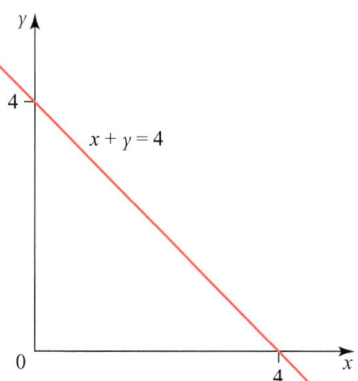

Figure 4.6

The coordinates of every point on that line give a pair of possible values for x and y. If the line $y = 2x + 1$ is also shown, as in Figure 4.7, the two lines can be seen to intersect at a single point.

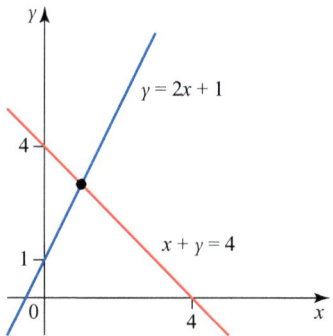

Figure 4.7

The coordinates of this point $(1, 3)$ are the solution $(x = 1, y = 3)$ to the *simultaneous equations*

$$x + y = 4$$

and $y = 2x + 1$.

There are several ways of solving simultaneous equations. You have just seen one method, that of drawing graphs. This is valid, but it has two drawbacks:

(i) it is tedious

(ii) it may not be very accurate, particularly if the solution does not have integer values.

Simultaneous equations in two unknowns

Solving simultaneous equations by substitution

Example 4.9

Solve the simultaneous equations

$$x + y = 4$$
$$y = 2x + 1$$

by substitution.

> This method is particularly suitable when one of the unknowns is already written as the subject of one of the equations.

Solution

Take the expression for y from the second equation and substitute it into the first. This gives

$$x + (2x + 1) = 4$$
$$\Rightarrow \quad 3x = 3$$
$$\Rightarrow \quad x = 1$$

Then substitute $x = 1$ into one of the original equations, e.g. $y = 2x + 1$, giving $y = 2 \times 1 + 1 = 3$.

So the solution is $x = 1$, $y = 3$, as indicated by the graphs.

Example 4.10

Figure 4.8 shows the graphs of $y = x^2 + x$ and $2x + y = 4$.

Solve the simultaneous equations

$$y = x^2 + x$$

and $\quad 2x + y = 4$

using the method of substitution.

> This method is also suitable when one of the equations represents a curve.

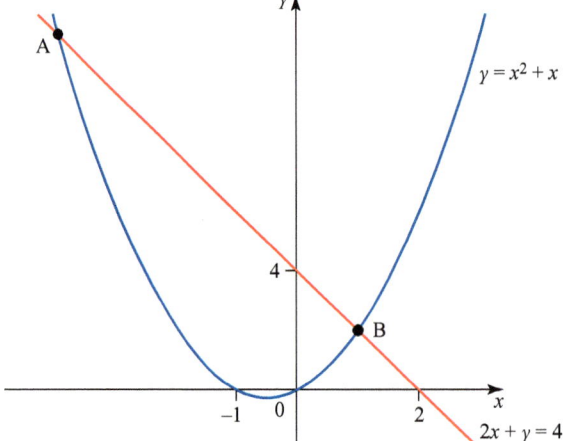

Figure 4.8

> **!** Notice from Figure 4.8 that there are two points of intersection, A and B, so expect the solution to be two pairs of values for x and y.

Solution

$$y = x^2 + x \quad \text{①}$$
$$2x + y = 4 \quad \text{②}$$

Substitute for y from equation ① into equation ②.

$$2x + (x^2 + x) = 4$$
$$\Rightarrow \quad x^2 + 3x - 4 = 0$$
$$\Rightarrow \quad (x + 4)(x - 1) = 0$$
$$\Rightarrow \quad x = -4 \text{ or } x = 1$$

Substituting into $2x + y = 4$: *(Always substitute back into the linear equation.)*

$x = -4 \Rightarrow -8 + y = 4 \Rightarrow y = 12$
$x = 1 \Rightarrow 2 + y = 4 \Rightarrow y = 2$

The solution is $x = -4$, $y = 12$ (point A) and $x = 1$, $y = 2$ (point B). *(Check your solution also fits equation ①.)*

> **Discussion point**
> → Having found the values of x in the example, the values of y were found by substituting into the equation of the line. Why was it advisable to use the linear equation rather than the quadratic?

> **!** The solution must always be given as pairs of values. It is wrong to write $x = -4$ or 1, $y = 12$ or 2, since not all pairs of values are possible.

Solving linear simultaneous equations by elimination

When both equations are linear and written in the same form, it may be preferable to use a process referred to as elimination.

Example 4.11

Solve the simultaneous equations

$$2x + y = 8 \quad \text{①}$$
$$5x + 2y = 21 \quad \text{②}$$

Solution

Notice that multiplying equation ① by 2 gives another equation containing $2y$.

$$5x + 2y = 21 \quad \text{equation ②}$$
$$\underline{4x + 2y = 16} \quad 2 \times \text{equation ①}$$
Subtracting $\Rightarrow \quad x = 5$

Substitute $x = 5$ into equation ①.

$$10 + y = 8 \quad \Rightarrow \quad y = -2$$

The solution is $x = 5$, $y = -2$.

Simultaneous equations in two unknowns

Sometimes you need to manipulate both equations to eliminate one of the unknowns, as in the following example.

Example 4.12

Solve the simultaneous equations

$$2x + 3y = -1 \quad \text{①}$$
$$3x - 2y = 18 \quad \text{②}$$

Solution

It is equally easy to eliminate x or y. It is up to you to choose which. The following method eliminates y.

$$4x + 6y = -2 \qquad 2 \times \text{equation ①}$$
$$9x - 6y = 54 \qquad 3 \times \text{equation ②}$$

Adding $\Rightarrow \quad 13x = 52$

$\Rightarrow \quad x = 4$

Substitute $x = 4$ into equation ①.

$8 + 3y = -1 \quad \Rightarrow \quad y = -3$

The solution is $x = 4, y = -3$.

> **Discussion point**
>
> → In Example 4.11 the equations were subtracted; in Example 4.12 they were added. How do you decide whether to add or subtract?

Simultaneous equations may arise in everyday problems.

Example 4.13

Tracey is buying fruit for a picnic.

Five apples and four pears cost exactly £2.20.

Two apples and six pears also cost exactly £2.20.

(i) Write this information as a pair of simultaneous equations.

(ii) Solve your equations to work out the cost of each type of fruit.

Solution

Let a pence be the cost of an apple and p pence be the cost of a pear.

> Make sure you introduce your unknowns.

> The cost of each piece of fruit will be a number of pence, so writing £2.20 as 220 pence avoids working with decimals.

(i) $5a + 4p = 220 \quad \text{①}$

$2a + 6p = 220 \quad \text{②}$

(ii) $\Rightarrow \quad 15a + 12p = 660 \qquad 3 \times \text{equation ①}$

$ 4a + 12p = 440 \qquad 2 \times \text{equation ②}$

Subtracting $\qquad 11a = 220$

$\Rightarrow \quad a = 20$

Substitute $a = 20$ into equation ①.

$100 + 4p = 220$

$\Rightarrow \quad p = 30$

An apple costs 20 pence and a pear costs 30 pence.

Example 4.14

A flag consists of a blue cross on a white background. Each white rectangle measures $2x$ cm by x cm, and the cross is y cm wide.

(i) Work out the total area of the flag in terms of x and y.

(ii) Show that the area of the cross is $6xy + y^2$.

(iii) The total area of the flag is 4500 cm² and the area of the cross is 1300 cm². Work out the values of x and y.

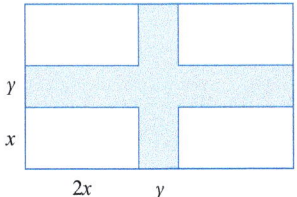

Figure 4.9

Solution

(i) Length $= 4x + y$ and width $= 2x + y$

so area $= (4x + y)(2x + y)$
$= 8x^2 + 6xy + y^2$

(ii) Each white rectangle has an area of $2x \times x = 2x^2$.

∴ area of cross $= 8x^2 + 6xy + y^2 - (4 \times 2x^2)$
$= 6xy + y^2$

(iii) $8x^2 + 6xy + y^2 = 4500$
$6xy + y^2 = 1300$

Subtracting $8x^2 \qquad = 3200$

$\Rightarrow \quad x^2 = 400$

$\Rightarrow \quad x = 20 \qquad$ (positive answer only)

Substitute $x = 20$ into $6xy + y^2 = 1300$

$120y + y^2 = 1300$

$\Rightarrow \quad y^2 + 120y - 1300 = 0$

$\Rightarrow \quad (y + 130)(y - 10) = 0$

$\Rightarrow \quad y = -130$ (reject since y is a length and so cannot be negative) or $y = 10$

$\Rightarrow \quad x = 20$ and $y = 10$

Exercise 4B

① Solve the following pairs of simultaneous equations using the substitution method.

(i) $y = x - 3$
$3x + 2y = 19$

(ii) $y = 2x - 9$
$4x - y = 17$

(iii) $y = 11 - 2x$
$2x + 5y = 37$

(iv) $y = 3x + 3$
$x - 2y = 4$

(v) $y = 7 - 2x$
$2x + 3y = 15$

(vi) $y = 3x - 5$
$x + 3y = -20$

Simultaneous equations in two unknowns

② Solve the following pairs of simultaneous equations using the elimination method.

(i) $3x + 2y = 12$ (ii) $3x - 2y = 6$ (iii) $3x + 2y = 22$
      $4x - y = 5$ $5x + 6y = 38$ $4x - 3y = 18$

(iv) $5x + 4y = 11$ (v) $4x + 5y = 33$ (vi) $4x - 3y = 2$
       $2x + 3y = 9$ $3x + 2y = 16$ $5x - 7y = 9$

③ Solve the following pairs of simultaneous equations.

(i) $x + y = 5$ (ii) $x - y + 1 = 0$ (iii) $x^2 + xy = 8$
     $x^2 + y^2 = 17$ $3x^2 - 4y = 0$ $x - y = 6$

(iv) $2x - y + 3 = 0$ (v) $x = 2y$ (vi) $x + 2y = -3$
       $y^2 - 5x^2 = 20$ $x^2 - y^2 + xy = 20$ $x^2 - 2x + 3y^2 = 11$

SA ④ For each of the following situations, form a pair of simultaneous equations and solve them to answer the question.

(i) Three chews and four lollipops cost 72p. Five chews and two lollipops cost 64p.

Work out the cost of a chew and the cost of a lollipop.

(ii) A taxi firm charges a fixed amount plus an extra fee per mile.

A journey of five miles costs £5.00 and a journey of seven miles costs £6.60.

How much does a journey of two miles cost?

(iii) Three packets of crisps and two packets of nuts cost £1.45. Two packets of crisps and five packets of nuts cost £2.25.

How much does one packet of crisps and four packets of nuts cost?

(iv) Two adults and one child paid £37.50 to go to the theatre. The cost for one adult and three children was also £37.50.

How much does it cost for two adults and five children?

SA ⑤ The diagram shows the circle $x^2 + y^2 = 25$ and the line $x + y = 7$.

Work out the coordinates of A and B.

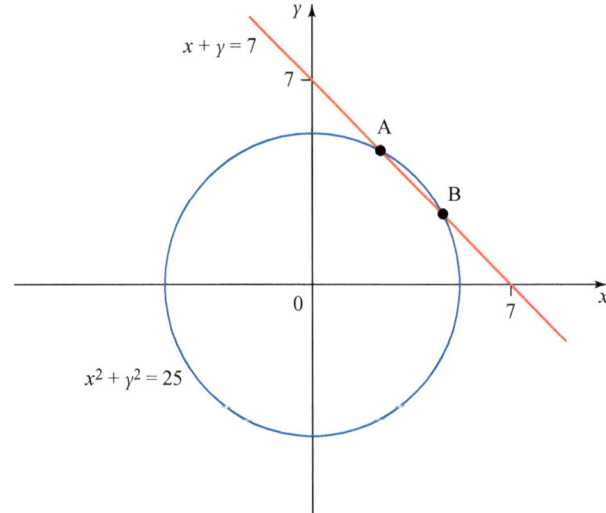

Figure 4.10

IR ⑥ The sum of two numbers is 10. The product is −96.

Work out the two numbers.

IR ⑦ (i) Work out the point of intersection of the circle $x^2 + y^2 = 8$ and the straight line $y - x = 4$.

(ii) There is only one point of intersection of the circle and the line in part (i). Which of these diagrams is a sketch of the two graphs?
Give a reason for your choice.

(a) 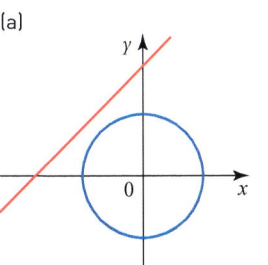   (b) 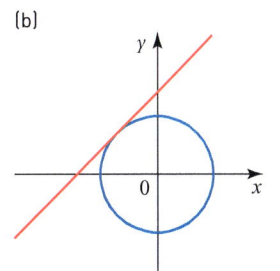   (c)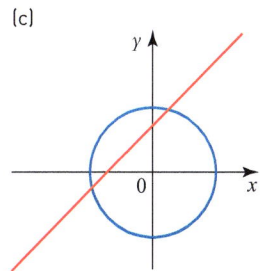

Figure 4.11

3 The remainder theorem

When a polynomial is divided by a linear expression then it will either divide exactly or there will be a remainder.

For example,

$$(x^2 + 3x + 2) \div (x + 1) = x + 2$$

> We know these results because $x^2 + 3x + 2 = (x + 1)(x + 2)$.

Also

$$(x^2 + 3x + 3) \div (x + 1) = x + 2 \text{ plus a remainder of } 1$$

In general, if P(x) is a polynomial, and Q(x) is the quotient when P(x) is divided by $(x - a)$ and r is the remainder, then

$$\frac{P(x)}{x - a} \equiv Q(x) + \frac{r}{x - a}$$

> A quotient is the result of a division. WJEC Additional Maths students will not be expected to be familiar with this word.

Multiplying throughout by $(x - a)$ gives

$$P(x) \equiv (x - a)Q(x) + r$$

This equivalence is true for all values of x.

When $x = a$ is substituted, then

$$P(a) \equiv (a - a)Q(a) + r \quad \Rightarrow \quad P(a) = 0 \times Q(a) + r \quad \Rightarrow \quad P(a) = r.$$

This is the *remainder theorem*.

> When a polynomial P(x) is divided by $(x - a)$, then the remainder will be P(a).

This can be extended to include division by a more general linear expression:

> When a polynomial P(x) is divided by $(ax + b)$, then the remainder will be $P\left(-\frac{b}{a}\right)$.

The factor theorem

Example 4.15
(i) Find the remainder when $(x^2 - 4x + 7)$ is divided by $(x + 6)$.
(ii) Find the remainder when $(x^3 + 5x - 7)$ is divided by $(2x - 1)$.

Solution

(i) Substitute $x = -6$ into $x^2 - 4x + 7$.

The remainder is $(-6)^2 - 4 \times (-6) + 7 = 36 + 24 + 7 = 67$.

(ii) Substitute $x = \frac{1}{2}$ into $x^3 + 5x - 7$.

The remainder is $\left(\frac{1}{2}\right)^3 + 5 \times \left(\frac{1}{2}\right) - 7 = \frac{1}{8} + \frac{5}{2} - 7 = -\frac{35}{8}$.

> This value of x is the solution of $2x - 1 = 0$.

Exercise 4C

① Find the remainder when $(x^2 + 4x - 9)$ is divided by $(x - 2)$.

② Find the remainders when each of these polynomials, P(x), is divided by the corresponding linear expression.

(i)	P(x) = $x^2 - 2x + 2$	($x - 5$)
(ii)	P(x) = $x^3 + x + 6$	($x + 1$)
(iii)	P(x) = $x^3 + x^2 - 7x - 6$	($x - 4$)
(iv)	P(x) = $x^2 - 3x - 2$	($2x - 1$)
(v)	P(t) = $t^4 - 2t^3 - 1$	($2t + 3$)
(vi)	P(c) = $c^5 + 3c^4 - 2c + 1$	($3c - 1$)

SA ③ When $(x^3 + ax - 3)$ is divided by $(x + 2)$ the remainder is 7.
Find the value of a.

SA ④ When $(x^4 - bx^2 + x - 2)$ is divided by $(x + 3)$ the remainder is -5.
Find the remainder when $(x^4 - bx^2 + x - 2)$ is divided by $(x - 2)$.

IR ⑤ P(x) = $x^3 + cx + d$

When P(x) is divided by $(x - 1)$ the remainder is -3.

When P(x) is divided by $(x + 2)$ the remainder is 3.

Find the values of c and d.

4 The factor theorem

The highest power in a quadratic is 2. Cubic expressions go up to 3, quartics to 4, quintics to 5, and so on. Such expressions are collectively referred to as polynomials. The degree of a polynomial is its highest power.

Note: a polynomial does not have negative or non-integer powers.

Just like the quadratic formula, there are formulae for solving cubic equations and quartic equations.

> **Prior knowledge**
> You should be familiar with function notation from your GCSE work.

> You should not attempt to learn this formula. It is included here for interest only.

The formula for solving the cubic equation $ax^3 + bx^2 + cx + d = 0$ is

$$x = \sqrt[3]{\left(\frac{-b^3}{27a^3} + \frac{bc}{6a^2} - \frac{d}{2a}\right) + \sqrt{\left(\frac{-b^3}{27a^3} + \frac{bc}{6a^2} - \frac{d}{2a}\right)^2 + \left(\frac{c}{3a} - \frac{b^2}{9a^2}\right)^3}}$$

$$+ \sqrt[3]{\left(\frac{-b^3}{27a^3} + \frac{bc}{6a^2} - \frac{d}{2a}\right) - \sqrt{\left(\frac{-b^3}{27a^3} + \frac{bc}{6a^2} - \frac{d}{2a}\right)^2 + \left(\frac{c}{3a} - \frac{b^2}{9a^2}\right)^3}} - \frac{b}{3a}$$

Clearly this is not a practical formula to use without a pre-programmed calculator, or a computer. The quartic formula is even more complicated, and too long to include here. Interestingly, it has been proved to be impossible to write a quintic formula.

In this course, you will only be asked to solve polynomial equations which can be reduced to linear and/or quadratic factors.

FUTURE USES

Some calculators can solve complicated equations. Such calculators use the Newton–Raphson method, which is a technique that students of Mathematics at A Level will learn. The above formula sometimes involves the use of imaginary numbers even if the final answers are not themselves imaginary. Students of A-Level Further Mathematics will learn about imaginary numbers (square roots of negative numbers).

Solving polynomial equations first involves use of the factor theorem.

Look at this quadratic equation.

$$x^2 - 5x - 6 = 0$$

Factorising $\Rightarrow$ $(x - 6)(x + 1) = 0$

$\Rightarrow$ $(x - 6) = 0$ or $(x + 1) = 0$

$\Rightarrow$ $x = 6$ or $x = -1$

> **Discussion points**
> → What happens if you substitute $x = 6$ into $x^2 - 5x - 6$?
> → What about $x = -1$?

The *factor theorem* states this result in a general form:

If $(x - a)$ is a factor of the polynomial f(x), then

- f(a) = 0
- $x = a$ is a root of the equation f(x) = 0.

Conversely, if f(a) = 0, then $(x - a)$ is a factor of f(x).

Example 4.16

Given that

f(x) = $x^3 + 2x^2 - x - 2$

(i) find f(1), f(−1), f(2), f(−2)

(ii) and hence factorise $x^3 + 2x^2 - x - 2$.

The factor theorem

Solution

(i) $f(1) = 1 + 2 - 1 - 2$
$= 0$ $\Rightarrow$ $(x - 1)$ is a factor.

$f(-1) = (-1)^3 + 2(-1)^2 - (-1) - 2$
$= -1 + 2 + 1 - 2$
$= 0$ $\Rightarrow$ $(x + 1)$ is a factor.

$f(2) = 8 + 8 - 2 - 2$
$= 12$ $\Rightarrow$ $(x - 2)$ is not a factor.

$f(-2) = (-2)^3 + 2(-2)^2 - (-2) - 2$
$= -8 + 8 + 2 - 2$
$= 0$ $\Rightarrow$ $(x + 2)$ is a factor.

(ii)   Hence $x^3 + 2x^2 - x - 2 = k(x - 1)(x + 1)(x + 2)$
where k is a constant.
The coefficient of x^3 is 1, so k must be 1.
$f(x) = (x - 1)(x + 1)(x + 2)$

Example 4.17

Given that
$f(x) = x^3 + 3x^2 - x - 3$

(i)   show that $(x + 1)$ is a factor of $f(x)$

(ii)   suggest other values of x you should try when looking for another factor

(iii)   solve the equation $f(x) = 0$.

Solution

(i) $f(-1) = (-1)^3 + 3(-1)^2 - (-1) - 3$
$= -1 + 3 + 1 - 3$
$= 0$
$\therefore (x + 1)$ is a factor of $f(x)$.

(ii)   Any other linear factor will be of the form $(x - a)$, where a is a factor of the constant term (-3).
This means that the only other values of x which are worth trying are 1, 3 and -3.

(iii) $f(1) = 1 + 3 - 1 - 3$
$= 0$ $\Rightarrow$ $(x - 1)$ is a factor.

$f(3) = 27 + 27 - 3 - 3$
$= 48$

$f(-3) = -27 + 27 + 3 - 3$
$= 0$ $\Rightarrow$ $(x + 3)$ is a factor.

As $f(x)$ is a cubic, then there are no more than three roots.
$x = -1, x = 1, x = -3$

Sometimes you may only be able to find one linear factor for the cubic and, in this case, you then need to use long division.

Example 4.18

Given that
$f(x) = x^3 - x^2 - 3x - 1$

(i) show that $(x + 1)$ is a factor
(ii) factorise $f(x)$
(iii) solve $f(x) = 0$.

Discussion points

→ What happens when you try $x = 1$?
→ Is there any other value you should try?

Solution

(i) $f(-1) = (-1)^3 - (-1)^2 - 3(-1) - 1$
$= -1 - 1 + 3 - 1$
$= 0$
$\Rightarrow (x + 1)$ is a factor of $x^3 - x^2 - 3x - 1$.

(ii) Since $(x + 1)$ is a factor, then divide $f(x)$ by $(x + 1)$.

$$\begin{array}{r}x^2 - 2x - 1\\ x+1{\overline{\smash{\big)}\,x^3 - x^2 - 3x - 1}}\\ \underline{x^3 + x^2}\\ -2x^2 - 3x\\ \underline{-2x^2 - 2x}\\ -x - 1\\ \underline{-x - 1}\\ 0\end{array}$$

$x^3 + x^2$ is $x^2 \times (x + 1)$

$-2x^2 - 2x$ is $-2x \times (x + 1)$

$-x - 1$ is $-1 \times (x + 1)$

$x^2 - 2x - 1$ cannot be factorised, so $f(x)$ is now fully factorised.

So $f(x) = (x + 1)(x^2 - 2x - 1)$.

(iii) $f(x) = 0 \Rightarrow (x + 1)(x^2 - 2x - 1) = 0$

$\Rightarrow$ either $x = -1$ or $x^2 - 2x - 1 = 0$.

Using the quadratic formula on $x^2 - 2x - 1 = 0$ gives

$x = \dfrac{2 \pm \sqrt{4 - (4 \times 1 \times (-1))}}{2}$

$= \dfrac{2 \pm \sqrt{8}}{2}$

$= 2.414$ or -0.414

The complete solution is $x = -1$, $x = -0.414$ or $x = 2.414$ (to 3 d.p.)

The factor theorem

> When using long division it is advisable to keep the terms in columns. This may mean that an extra zero term should be included to help with this.

Example 4.19

Given that $(x + 2)$ is a factor of $x^3 - 5x - 2$, work out a quadratic factor.

Solution

Method 1

$$
\begin{array}{r}
x^2 - 2x - 1 \\
x + 2 \overline{\smash{)}\, x^3 + 0x^2 - 5x - 2} \\
\underline{x^3 + 2x^2 } \\
-2x^2 - 5x \\
\underline{-2x^2 - 4x } \\
-x - 2 \\
\underline{-x - 2} \\
0
\end{array}
$$

$\Rightarrow \quad x^2 - 2x - 1$ is also a factor.

Method 2

As $(x + 2)$ is a factor, $x^3 - 5x - 2 \equiv (x + 2)(ax^2 + bx + c)$

You could compare coefficients as shown in Chapter 2.

However, it is clear that a must be 1 and c must be -1, so it is simpler to immediately write

$x^3 - 5x - 2 \equiv (x + 2)(x^2 + bx - 1)$

and then compare coefficients of x^2 or x to find b:

$0 = 2 + b$ or $-5 = 2b - 1$ both of which give $b = -2$,

so $x^3 - 5x - 2 \equiv (x + 2)(x^2 - 2x - 1)$.

An extension to the factor theorem includes simplified fractional roots.

$$f\left(\frac{b}{a}\right) = 0 \iff (ax - b) \text{ is a factor of } f(x).$$

Example 4.20

(i) Show that $(x + 1)$ and $(3x - 2)$ are factors of
$3x^4 + 4x^3 - 16x^2 - 7x + 10$.

(ii) Hence solve $3x^4 + 4x^3 - 16x^2 - 7x + 10 = 0$.

Solution

(i) Let $f(x) = 3x^4 + 4x^3 - 16x^2 - 7x + 10$.

$f(-1) = 3 \times (-1)^4 + 4 \times (-1)^3 - 16 \times (-1)^2 - 7 \times (-1) + 10$

$\qquad = 3 - 4 - 16 + 7 + 10$

$\qquad = 0$

$\Rightarrow \quad (x + 1)$ is a factor of $3x^4 + 4x^3 - 16x^2 - 7x + 10$.

$f\left(\dfrac{2}{3}\right) = 3 \times \left(\dfrac{2}{3}\right)^4 + 4 \times \left(\dfrac{2}{3}\right)^3 - 16 \times \left(\dfrac{2}{3}\right)^2 - 7 \times \dfrac{2}{3} + 10$

$\qquad = \dfrac{16}{27} + \dfrac{32}{27} - \dfrac{64}{9} - \dfrac{14}{3} + 10$

$\qquad = 0$

$\Rightarrow \quad (3x - 2)$ is also a factor of $3x^4 + 4x^3 - 16x^2 - 7x + 10$.

(ii) $(x + 1)(3x - 2) = 3x^2 + x - 2$

$$
\begin{array}{r}
x^2 + x - 5 \\
3x^2 + x - 2 \overline{\smash{\big)}\, 3x^4 + 4x^3 - 16x^2 - 7x + 10} \\
\underline{3x^4 + x^3 - 2x^2 } \\
3x^3 - 14x^2 - 7x \\
\underline{3x^3 + x^2 - 2x } \\
-15x^2 - 5x + 10 \\
\underline{-15x^2 - 5x + 10} \\
0
\end{array}
$$

$\Rightarrow \quad f(x) = (x + 1)(3x - 2)(x^2 + x - 5)$

This equals zero when

$x + 1 = 0, \; 3x - 2 = 0, \; x^2 + x - 5 = 0$

$\Rightarrow x = -1 \text{ or } x = \dfrac{2}{3} \quad \text{or} \quad x = \dfrac{-1 \pm \sqrt{1^2 - 4 \times 1 \times -5}}{2 \times 1}$

$\qquad\qquad\qquad\qquad\qquad\qquad\qquad \Rightarrow x = \dfrac{-1 \pm \sqrt{21}}{2}$

Exercise 4D

① Determine whether the following linear functions are factors of the given polynomials.

(i) $x^3 - 8x + 7$ $(x - 1)$

(ii) $x^3 + x^2 - 4x - 5$ $(x + 2)$

(iii) $x^4 - 6x^2 + 10x - 12$ $(x - 2)$

(iv) $x^5 + 32$ $(x + 2)$

(v) $2x^4 - x^3 - 20$ $(x + 2)$

(vi) $x^3 - ax^2 + a^2x - a^3$ $(x - a)$

Algebraic proof

> Exam questions are unlikely to require you to find four or five different linear factors of an expression. However, you will be expected to feel comfortable working with polynomials of such a high degree.

② Factorise the following functions as a product of linear factors.
- (i) $x^3 - 3x^2 - x + 3$
- (ii) $x^3 - 7x - 6$
- (iii) $x^3 - x^2 - 2x$
- (iv) $x^3 - 2x^2 - 13x - 10$
- (v) $x^3 - x^2 - 14x + 24$
- (vi) $x^4 - 3x^3 - 11x^2 + 3x + 10$
- (vii) $x^4 - 4x^3 + 6x^2 - 4x + 1$
- (viii) $x^4 - 13x^2 + 36$
- (ix) $x^5 - 4x^4 - 17x^3 + 24x^2 + 36x$
- (x) $x^5 - 3x^4 - 23x^3 + 51x^2 + 94x - 120$

③ Solve the following equations.
- (i) $x^3 - 2x^2 - 5x + 6 = 0$
- (ii) $x^3 + 3x^2 - 6x - 8 = 0$
- (iii) $x^3 - 2x^2 - 21x - 18 = 0$
- (iv) $x^4 + 3x^3 - 5x^2 - 3x + 4 = 0$
- (v) $2x^3 + x^2 - 7x + 4 = 0$
- (vi) $x^5 - 3x^4 - 23x^3 + 51x^2 + 94x - 120 = 0$

④ $f(x) = x^3 + 2x^2 + ax - 76$

Given that $(x - 4)$ is a factor of $f(x)$, work out the value of a.

⑤ $f(x) = x^3 + px^2 + qx + 6$
- (i) Given that $(x - 1)$ is a factor of $f(x)$, write an equation in p and q.
- (ii) Given also that $(x + 3)$ is a factor of $f(x)$, write another equation in p and q.
- (iii) Solve your simultaneous equations to work out the values of p and q.

 ⑥ (i) Work out the value of k for which $x = 2$ is a root of $x^3 + kx + 6 = 0$.
- (ii) Work out the other roots when k takes this value.

 ⑦ The diagram shows an open cuboid tank whose base is a square of side x metres and whose volume is 8 m³.
- (i) Write down an expression in terms of x for the height of the tank.
- (ii) Show that the surface area of the tank is $\left(x^2 + \dfrac{32}{x}\right)$ m².
- (iii) Given that the surface area is 24 m², show that
$x^3 - 24x + 32 = 0.$
- (iv) Solve $x^3 - 24x + 32 = 0$ to work out the possible values of x.

Figure 4.12

 ⑧ $(x - 1)$ and $(5x + 2)$ are factors of $5x^4 + px^3 - 10x^2 + qx + 2$.

Solve $5x^4 + px^3 - 10x^2 + qx + 2 = 0$.

5 Algebraic proof

Any of the algebraic skills covered in previous sections may be needed in proofs.

When constructing a proof, avoid writing the required result as the first line of working. This is a common error. Instead, start with a given expression and gradually change it using algebraic processes, or start with a known fact which, when combined with other known facts, proves the required result.

Example 4.21

Prove that $2a^3 - a^2(2a - 9)$ is a square number when a is an integer.

Solution

Expand and simplify
$$2a^3 - a^2(2a - 9) = 2a^3 - 2a^3 + 9a^2$$
$$= 9a^2$$
$$= (3a)^2$$

As a is an integer then $3a$ is also an integer.

$\therefore$ $2a^3 - a^2(2a - 9)$ is a square number when a is an integer.

> The final line of any proof should be a repeat of the required statement in the question.

Example 4.22

(i) Express $x^2 - 8x + 18$ in the form $(x - a)^2 + b$ where a and b are integers.

(ii) Hence, prove that $x^2 - 8x + 18$ is always positive.

Solution

(i) $x^2 - 8x + 18 \equiv (x - a)^2 + b$

$x^2 - 8x + 18 \equiv x^2 - 2ax + a^2 + b$

Equate coefficients of x $\quad -8 = -2a$

$\quad\quad\quad\quad\quad\quad\quad\quad\quad\quad\quad 4 = a$

Equate constants $\quad\quad 18 = a^2 + b$

$\quad\quad\quad\quad\quad\quad\quad\quad\quad 18 = 16 + b$

$\quad\quad\quad\quad\quad\quad\quad\quad\quad\quad 2 = b$

$x^2 - 8x + 18 \equiv (x - 4)^2 + 2$

> An alternative is to complete the square on $x^2 - 8x$,
> i.e. $x^2 - 8x + 18$
> $= (x - 4)^2 - 4^2 + 18$
> $= (x - 4)^2 + 2$.

(ii) You know that $(x - 4)^2 \geq 0$

$\Rightarrow (x - 4)^2 + 2 \geq 2$

$\Rightarrow (x - 4)^2 + 2 > 0$

$\Rightarrow x^2 - 8x + 18 > 0$

> When a number is squared the answer can never be negative.

Example 4.23

c and d are positive integers such that $c > d$.

$$f(x) = \frac{2c + cx}{2d + dx} \quad x \neq -2$$

Prove that $f(x) > 1$.

Solution

Factorise the numerator and denominator $\quad f(x) = \dfrac{c(2 + x)}{d(2 + x)}$

Cancel $(2 + x)$ $\quad\quad\quad\quad\quad\quad\quad\quad\quad\quad f(x) = \dfrac{c}{d}$

But $c > d$ (and d is positive) $\Rightarrow \dfrac{c}{d} > 1 \quad \therefore \; f(x) > 1.$

Algebraic proof

Exercise 4E

1. Prove that $2(m + 7) - 2(5 + m)$ is always a positive integer.
2. Prove that $5(c - 3) + 3(c + 7)$ is always even when c is a positive integer.
3. Prove that $(y + 6)(y + 3) - y^2$ is a multiple of 9 when y is a positive integer.
4. Prove the following identities.

   (i) $\dfrac{x}{2} + \dfrac{x - 1}{3} \equiv \dfrac{5x - 2}{6}$

   (ii) $\dfrac{2x}{3} + \dfrac{x + 1}{4} \equiv \dfrac{11x + 3}{12}$

   (iii) $\dfrac{y + 1}{4} - \dfrac{y}{6} \equiv \dfrac{y + 3}{12}$

   (iv) $\dfrac{y}{2} + \dfrac{y - 1}{3} + \dfrac{y + 1}{4} \equiv \dfrac{13y - 1}{12}$

   (v) $\dfrac{5}{x} - \dfrac{x - 1}{3} \equiv \dfrac{15 - x^2 + x}{3x}$

   (vi) $\dfrac{x + 1}{6} + \dfrac{2x - 3}{9} - \dfrac{3x - 2}{12} \equiv \dfrac{5x}{36}$

5. $f(n) = n^2$ for all positive integer values of n.

   (i) Show that $f(n + 1) = n^2 + 2n + 1$.

   (ii) Prove that $f(n + 1) + f(n - 1)$ is always even.

   (iii) Prove that $f(n + 1) - f(n - 1)$ is always a multiple of 4.

6. (i) Express $x^2 + 2x + 5$ in the form $(x + a)^2 + b$ where a and b are integers.

   (ii) Hence, prove that $x^2 + 2x + 5$ is always positive.

7. Prove that $y^2 - 10y + 26 > 0$ for all values of y.

8. Prove that $9m^2(3m - 1) + (3m)^2$ is a cube number when m is a positive integer.

9. Prove that $\dfrac{6p - 18}{2p - 6}$ is always a positive integer when $p \neq 3$.

10. a is a positive number, b is a negative number.

    $a \neq -b$

    Prove that $\dfrac{a^2 + ab}{ab + b^2}$ is negative.

11. $f(x) = x^2 + 2x$.

    Prove that $f(4x) = kx(2x + 1)$ where k is an integer.

12. Four sides of a hexagon are $(2x + 8)$ cm, $(2y - 4)$ cm, $(x - 2)$ cm and $(y + 1)$ cm as shown.

    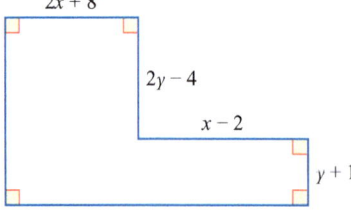

    Figure 4.13

    The perimeter of the hexagon is 42 cm.

    (i) Show that $x + y = 6$.

    (ii) Hence show that the area is $(-7x^2 + 15x + a)$ cm^2, where a is a value to be found.

FUTURE USES

Students who go on to study mathematics at A-Level will learn a variety of proof techniques, including proof by contradiction and proof by induction.

LEARNING OUTCOMES

Now you have finished this chapter, you should be able to
- solve quadratic equations
  - by factorising
  - by completing the square
  - using the quadratic formula
  - by drawing a graph
- solve simultaneous equations
  - in two unknowns
  - by plotting their graphs
- use the remainder theorem
  - to find the remainder when a polynomial is divided by a linear expression
- use the factor theorem
  - to factorise a polynomial
  - to solve a polynomial equation
- prove mathematical statements algebraically.

KEY POINTS

1. When factorising quadratics of the form $ax^2 + bx + c$, find two numbers with a sum of b and product ac. Then split the coefficient of x into these two numbers.
2. When completing the square on a quadratic of the form $ax^2 + bx + c$, take a out as a factor, or divide both sides of the equation by a.
3. If $ax^2 + bx + c = 0$ then $x = \dfrac{-b \pm \sqrt{b^2 - 4ac}}{2a}$
4. $P\left(-\dfrac{b}{a}\right)$ is the remainder when polynomial $P(x)$ is divided by $(ax + b)$.
5. $f(a) = 0 \iff (x - a)$ is a factor of $f(x)$.
6. In a proof question, show every step clearly.

5 Coordinate geometry

> Most of the fundamental ideas of science are essentially simple and may, as a rule, be expressed in a language comprehensible to everyone.
>
> Albert Einstein

1 Parallel and perpendicular lines

Prior knowledge

In Chapter 3 we used this fact:

The line joining (x_1, y_1) to (x_2, y_2) has gradient m, where $m = \dfrac{y_2 - y_1}{x_2 - x_1}$.

If you know the gradients m_1 and m_2 of two lines, you can tell at once if they are parallel or perpendicular.

Parallel lines: $m_1 = m_2$ Perpendicular lines: $m_1 m_2 = -1$

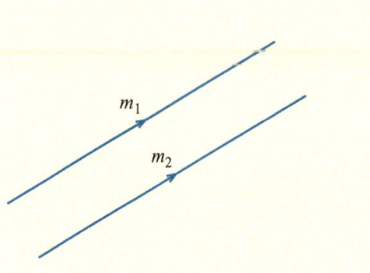
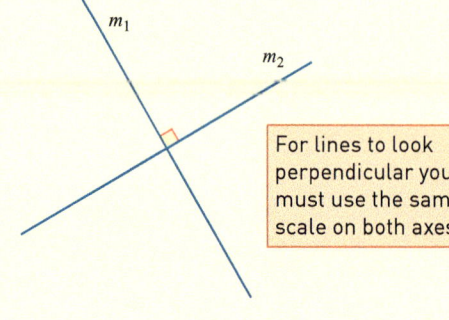

For lines to look perpendicular you must use the same scale on both axes.

Figure 5.1

To illustrate the result for perpendicular lines, try Activity 5.1 on squared paper.

Discussion point
→ How would you explain the result for parallel lines?

ACTIVITY 5.1

(i) Draw two congruent right-angled triangles in the positions shown in Figure 5.2. p and q can take any values.

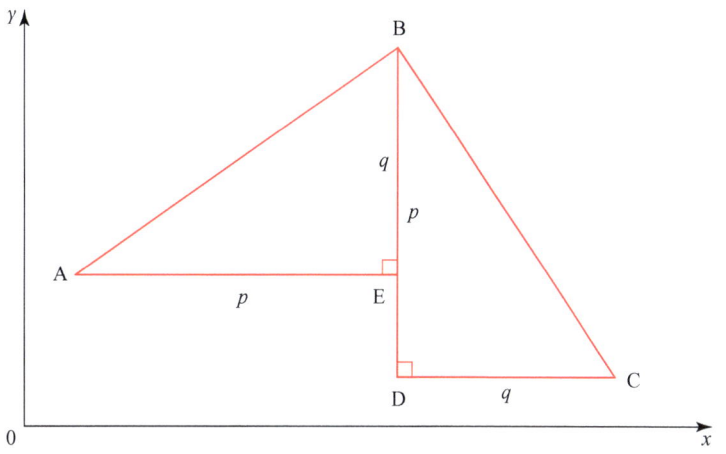

Figure 5.2

> ⚠ The gradient of a line perpendicular to a line of gradient m is given by $-\dfrac{1}{m}$.
> Don't forget to change the sign when taking the reciprocal.

(ii) Explain why $\angle ABC = 90°$.
(iii) Calculate the gradient of AB (m_1) and the gradient of BC (m_2).
(iv) Show that $m_1 m_2 = -1$.

2 The distance between two points

Look at Figure 5.3. P is (3, 1) and Q is (6, 5).

Prior knowledge

You are expected to use Pythagoras' theorem to calculate the distance between two points with known coordinates.

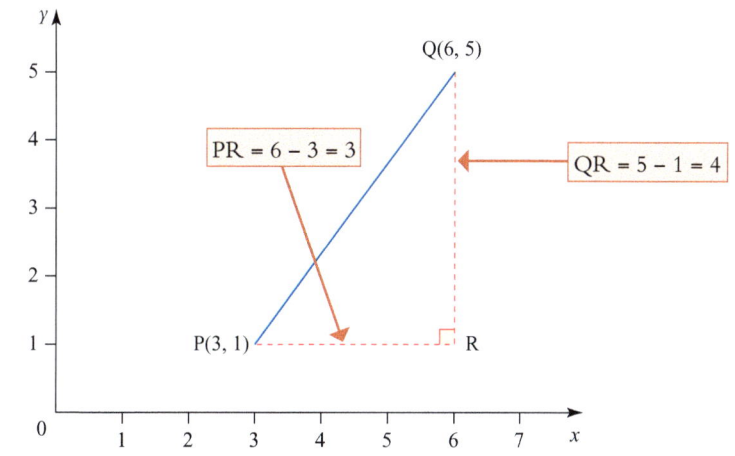

Figure 5.3

$$PQ = \sqrt{3^2 + 4^2} = \sqrt{25} = 5$$

Generalising this, if P has coordinates (x_1, y_1) and Q has coordinates (x_2, y_2), then length $PQ = \sqrt{(x_2 - x_1)^2 + (y_2 - y_1)^2}$.

3 The midpoint of a line joining two points

Look at the line joining the points P(1, 2) and Q(7, 4) in Figure 5.4. The point M is the midpoint of PQ.

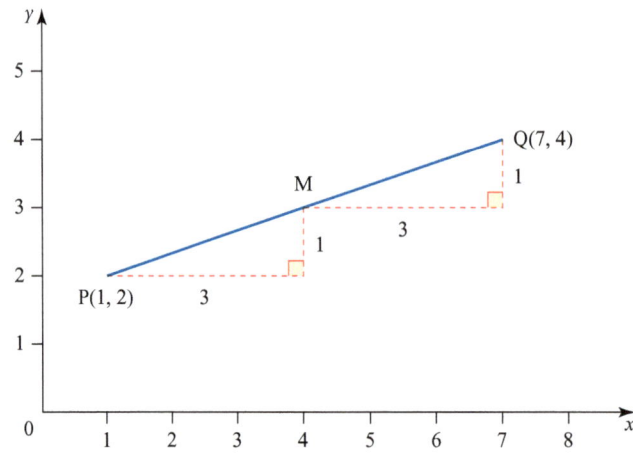

Figure 5.4

The coordinates of M are the means (averages) of the coordinates of P and Q.

$$\tfrac{1}{2}(1 + 7) = 4 \text{ and } \tfrac{1}{2}(2 + 4) = 3$$

M is $(4, 3)$.

Again, if P has coordinates (x_1, y_1) and Q has coordinates (x_2, y_2), then the coordinates of the midpoint of PQ are given by

$$\text{midpoint} = \left(\frac{x_1 + x_2}{2}, \frac{y_1 + y_2}{2} \right).$$

Example 5.1

A and B are the points $(-4, 2)$ and $(2, 5)$. Work out

(i)   the gradient of AB

(ii) the gradient of the line perpendicular to AB

(iii) the length of AB

(iv) the coordinates of the midpoint of AB.

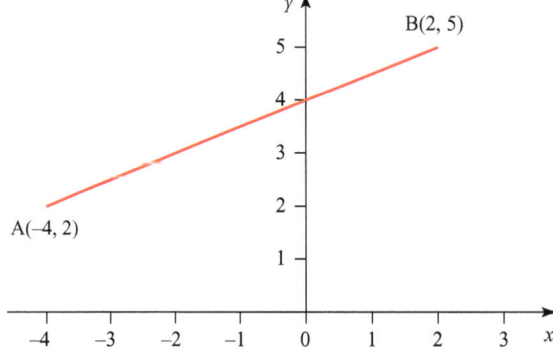

Figure 5.5

Solution

(i) Taking $(-4, 2)$ as (x_1, y_1) and $(2, 5)$ as (x_2, y_2)

$$\text{gradient} = \frac{5-2}{2-(-4)} = \frac{3}{6} = \frac{1}{2}.$$

(ii) $m_1 = \frac{1}{2}$ and $m_1 m_2 = -1$

$$\Rightarrow \quad \frac{1}{2} m_2 = -1$$

$$\Rightarrow \quad m_2 = -2$$

The line perpendicular to AB has gradient -2.

(iii) $\text{length} = \sqrt{(2-(-4))^2 + (5-2)^2}$

$= \sqrt{36 + 9}$

$= \sqrt{45}$

$= 6.71 \text{ (3 s.f.)}$

(iv) $\text{midpoint} = \left(\frac{-4+2}{2}, \frac{2+5}{2} \right)$

$= (-1, 3.5)$

Example 5.2

P is the point (a, b) and Q is the point $(3a, 5b)$.

Write expressions, in terms of a and b, for

(i) the gradient of PQ

(ii) the length of PQ

(iii) the midpoint of PQ.

Solution

Taking (a, b) as (x_1, y_1) and $(3a, 5b)$ as (x_2, y_2)

(i) $\text{gradient} = \frac{5b - b}{3a - a}$

$= \frac{4b}{2a}$

$= \frac{2b}{a}$

(ii) $\text{length} = \sqrt{(3a-a)^2 + (5b-b)^2}$

$= \sqrt{4a^2 + 16b^2}$

(iii) $\text{midpoint} = \left(\frac{a+3a}{2}, \frac{b+5b}{2} \right)$

$= (2a, 3b)$

> **Discussion point**
>
> → How can the length in part (ii) be simplified further?

The midpoint of a line joining two points

Example 5.3

A, B and C are the points $(1, 2)$, $(5, b)$ and $(6, 2)$. $\angle ABC = 90°$.

(i) Work out two possible values of b.

(ii) Show all four points on a sketch and describe the shape of the figure you have drawn.

Solution

(i) Gradient of AB $= \dfrac{b-2}{5-1} = \dfrac{b-2}{4}$

Gradient of BC $= \dfrac{2-b}{6-5} = 2 - b$

$\angle ABC = 90° \Rightarrow$ AB and BC are perpendicular.

$\Rightarrow \left(\dfrac{b-2}{4}\right) \times (2 - b) = -1$

$\Rightarrow (b-2)(2-b) = -4$

$\Rightarrow 2b - b^2 - 4 + 2b = -4$

$\Rightarrow 4b - b^2 = 0$

$\Rightarrow b(4 - b) = 0$

So $b = 0$ or $b = 4$.

(ii)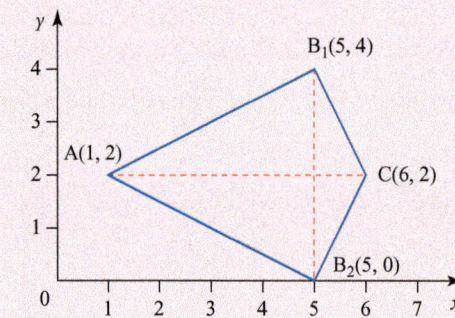

Figure 5.6

AB_1CB_2 is a quadrilateral with diagonals that are perpendicular, since AC is parallel to the x-axis and B_1B_2 is parallel to the y-axis.

This makes AB_1CB_2 a kite.

Exercise 5A

1. For each of the following pairs of points A and B, calculate
   (a) the gradient of the line perpendicular to AB
   (b) the length of AB
   (c) the coordinates of the midpoint of AB.

   (i) A(4, 3)   B(8, 11)

   (ii) A(3, 4)   B(0, 13)

   (iii) A(5, 3)   B(10, −8)

(iv)	A(−6, −14)	B(1, 7)
(v)	A(6, 0)	B(8, 15)
(vi)	A(−2, −4)	B(3, 9)
(vii)	A(−3, −6)	B(2, −7)
(viii)	A(4, 7)	B(7, −4)

② A(0, 5), B(4, 1) and C(2, 7) are the vertices of a triangle. Show that the triangle is right-angled
   (i) by finding the gradients of the sides
   (ii) by finding the lengths of the sides.

③ A(3, 6), B(7, 4) and C(1, 2) are the vertices of a triangle. Show that ABC is a right-angled isosceles triangle.

④ A(3, 5), B(3, 11) and C(6, 2) are vertices of a triangle.
   (i) Work out the perimeter of the triangle.
   (ii) Using AB as the base, work out the area of the triangle.

⑤ A quadrilateral PQRS has vertices at P(−2, −5), Q(11, −7), R(9, 6) and S(−4, 8).
   (i) Work out the lengths of the four sides of PQRS.
   (ii) Work out the midpoints of the diagonals PR and QS.
   (iii) Without drawing a diagram, show why PQRS cannot be a square. What is it?

⑥ The points A, B and C have coordinates (2, 3), (6, 12) and (11, 7) respectively.
   (i) Draw the triangle ABC.
   (ii) Show by calculation that the triangle is isosceles and name the two equal sides.
   (iii) Work out the midpoint of the third side.
   (iv) By calculating appropriate lengths, work out the area of triangle ABC.

⑦ A parallelogram WXYZ has three of its vertices at W(2, 1), X(−1, 5) and Y(−3, 3).
   (i) Work out the midpoint of WY.
   (ii) Use this information to work out the coordinates of Z.

⑧ A triangle ABC has vertices at A(3, 2), B(4, 0) and C(8, 2).
   (i) Show that the triangle is right-angled.
   (ii) Work out the coordinates of the point D such that ABCD is a rectangle.

⑨ The three points P(−2, 3), Q(1, q) and R(7, 0) are collinear (i.e. they lie on the same straight line).
   (i) Work out the value of q.
   (ii) Work out the ratio of the lengths PQ : QR.

⑩ A quadrilateral has vertices A(−2, 8), B(−5, 5), C(5, 3) and D(3, 7).
   (i) Draw the quadrilateral.
   (ii) Show by calculation that it is a trapezium.
   (iii) Work out the coordinates of E when ABCE is a parallelogram.

Equation of a straight line

4 Equation of a straight line

Prior knowledge

In Chapter 3 you used these three facts.
- The equation of the line with gradient m cutting the y-axis at the point $(0, c)$ is $y = mx + c$.
- The equation of the line with gradient m passing through (x_1, y_1) is $y - y_1 = m(x - x_1)$.
- The equation of the line passing through (x_1, y_1) and (x_2, y_2) is
$$\frac{y - y_1}{x - x_1} = \frac{y_2 - y_1}{x_2 - x_1}.$$

Example 5.4

An isosceles triangle with AB = AC has vertices at A(2, 3), B(8, 5) and C(4, 9).

Work out the equation of the line of symmetry.

Solution

Figure 5.7 shows the triangle ABC, with the line of symmetry joining A to the midpoint of BC.

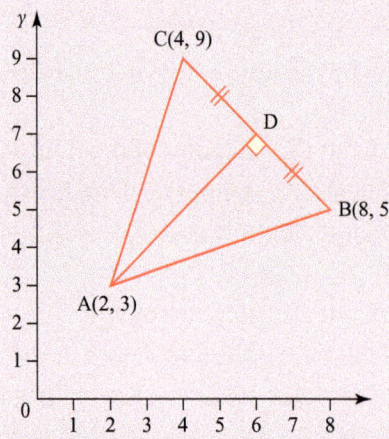

Figure 5.7

The coordinates of D are $\left(\dfrac{8 + 4}{2}, \dfrac{5 + 9}{2}\right) = (6, 7)$.

Let (x_1, y_1) be $(2, 3)$ and (x_2, y_2) be $(6, 7)$.

$$\frac{y - y_1}{y_2 - y_1} = \frac{x - x_1}{x_2 - x_1}$$

$$\Rightarrow \quad \frac{y - 3}{7 - 3} = \frac{x - 2}{6 - 2}$$

$$\Rightarrow \quad \frac{y - 3}{4} = \frac{x - 2}{4}$$

$$\Rightarrow \quad y = x + 1$$

Example 5.5

The straight line with equation $5x - 4y = 40$ intersects the x-axis at P and the y-axis at Q.

(i) Work out the area of triangle OPQ where O is the origin.

(ii) Work out the equation of the line that passes through Q and is perpendicular to PQ.

Solution

(i) Work out the coordinates of P and Q.

Substitute $y = 0$ in equation of line $\quad 5x - 0 = 40$
$$x = 8 \quad P(8, 0)$$

Substitute $x = 0$ in equation of line $\quad 0 - 4y = 40$
$$y = -10 \quad Q(0, -10)$$

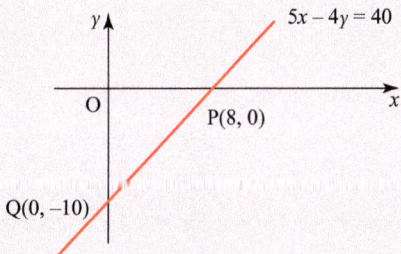

A sketch graph will often be useful.

Figure 5.8

Distance OP = 8 and distance OQ = 10.

$$\text{Area of triangle} = \tfrac{1}{2} \times \text{base} \times \text{height}$$
$$= \tfrac{1}{2} \times 8 \times 10$$
$$= 40 \text{ units}^2$$

(ii) Work out the gradient of PQ.

$$\frac{0 - (-10)}{8 - 0} = \frac{10}{8}$$
$$= \frac{5}{4}$$

Gradient of line perpendicular to PQ $= -\tfrac{4}{5}$

Line passes through $(0, -10)$ $\quad y = -\tfrac{4}{5}x - 10$

Equation of a straight line

Exercise 5B

① By calculating the gradients of the following pairs of lines, state whether they are parallel, perpendicular or neither.

(i) $x = 2$
       $y = -2$

(ii) $y = 2x$
       $y = -2x$

(iii) $x + 2y = 1$
       $2x - y = 1$

(iv) $y = x - 3$
       $x - y + 4 = 0$

(v) $y = 3 - 4x$
       $y = 4 - 3x$

(vi) $x + y = 5$
       $x - y = 5$

(vii) $x - 2y = 3$
       $y = \frac{1}{2}x - 1$

(viii) $x + 3y - 4 = 0$
       $y = 3x + 4$

(ix) $2y = x$
       $2x + y = 4$

(x) $2x + 3y - 4 = 0$
       $2x + 3y - 6 = 0$

(xi) $x + 3y = 1$
       $y + 3x = 1$

(xii) $2x = 5y$
       $5x + 2y = 0$

② Work out the equations of these lines.
(i) Parallel to $y = 3x$ and passing through $(3, -1)$
(ii) Parallel to $y = 2x + 3$ and passing through $(0, 7)$
(iii) Parallel to $y = 3x - 4$ and passing through $(3, -7)$
(iv) Parallel to $4x - y + 2 = 0$ and passing through $(5, 0)$
(v) Parallel to $3x + 2y - 1 = 0$ and passing through $(3, -2)$
(vi) Parallel to $2x + 4y - 5 = 0$ and passing through $(0, 5)$

③ Work out the equations of these lines.
(i) Perpendicular to $y = 2x$ and passing through $(0, 0)$
(ii) Perpendicular to $y = 3x - 1$ and passing through $(0, 4)$
(iii) Perpendicular to $y + x = 2$ and passing through $(3, -1)$
(iv) Perpendicular to $2x - y + 4 = 0$ and passing through $(1, -1)$
(v) Perpendicular to $3x + 2y + 4 = 0$ and passing through $(3, 0)$
(vi) Perpendicular to $2x + y - 1 = 0$ and passing through $(4, 1)$

④ Points P and Q have coordinates P(3, −1) and Q(5, 7).
(i) Work out the gradient of PQ.
(ii) Work out the coordinates of the midpoint of PQ.
(iii) The perpendicular bisector of a line PQ is the line which is perpendicular to PQ and passes through its midpoint.
Work out the equation of the perpendicular bisector of PQ.

⑤ A triangle has vertices P(2, 5), Q(−2, −2) and R(6, 0).
(i) Sketch the triangle.
(ii) Work out the coordinates of L, M and N, which are the midpoints of PQ, QR and RP respectively.
(iii) Work out the equations of the lines LR, MP and NQ (these are the medians of the triangle).
(iv) Show that the point (2, 1) lies on all three of these lines. (This shows that the medians of a triangle are concurrent.)

⑥ The straight line with equation $2x + 3y - 12 = 0$ cuts the x-axis at A and the y-axis at B.
  (i) Sketch the line.
  (ii) Work out the coordinates of A and B.
  (iii) Work out the area of triangle OAB where O is the origin.
  (iv) Work out the equation of the line which passes through O and is perpendicular to AB.
  (v) Work out the length of AB and, using the result in (iii), calculate the shortest distance from O to AB.

⑦ A quadrilateral has vertices at the points A(−7, 0), B(2, 3), C(5, 0) and D(−1, −6).
  (i) Sketch the quadrilateral.
  (ii) Work out the gradient of each side.
  (iii) Work out the equation of each side.
  (iv) Work out the length of each side.
  (v) Work out the area of the quadrilateral.

⑧ £10 000 is invested and simple interest of 2% per annum is received on this investment. (Simple interest is when the interest received each year is calculated on the initial investment in the account only.)
  (i) Calculate the interest received after each of the first three years.
  (ii) Sketch the graph of interest against time and write down its equation.
  (iii) Use the equation to work out how long it would take for the investment to reach £11 000.

⑨ A spring has an unstretched length (often called the natural length) of 20 cm. When it is hung with a load of 50 g attached, its stretched length is 25 cm.

Assuming that the extension of the spring is proportional to the load at all times
  (i) calculate the load corresponding to an extension of 12.5 cm
  (ii) calculate the extension corresponding to a load of 75 g
  (iii) calculate the extension corresponding to a load of 800 g and comment on your answer.

5 The intersection of two lines

Prior knowledge
You learned to solve simultaneous linear equations in Chapter 4.

You can work out the point of intersection of any two lines (or curves) by solving their equations simultaneously.

Example 5.6
(i) Sketch the lines $x + 3y - 6 = 0$ and $y = 2x - 5$ on the same axes.
(ii) Work out the coordinates of the point where they intersect.

The intersection of two lines

Solution

(i) The line $x + 3y - 6 = 0$ passes through $(0, 2)$ and $(6, 0)$.

The line $y = 2x - 5$ passes through $(0, -5)$ and has a gradient of 2.

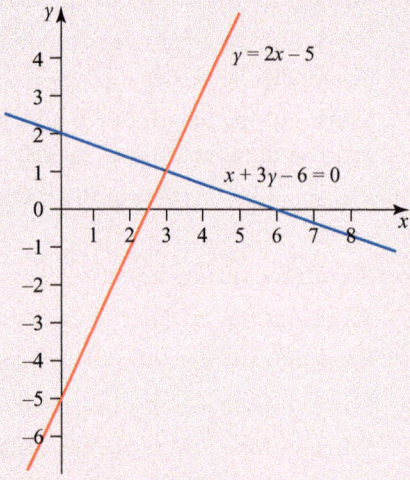

Figure 5.9

(ii) $x + 3y - 6 = 0 \Rightarrow 2x + 6y - 12 = 0$ (multiplying by 2) ①

$y = 2x - 5 \Rightarrow 2x - y - 5 = 0$ ②

① − ② $\Rightarrow 7y - 7 = 0$

$\Rightarrow y = 1$

Substituting $y = 1$ in ① gives $2x + 6 - 12 = 0$

$\Rightarrow x = 3$

The coordinates of the point of intersection are therefore $(3, 1)$.

> **Discussion point**
> → Graphical methods such as this will have limited accuracy. What factors would affect the accuracy of your solution in this case?

An alternative method for solving these equations simultaneously would be to plot both lines on graph paper and read off the coordinates of the point of intersection.

Example 5.7

(i) Plot the lines $x + y - 2 = 0$ and $4y - x = 4$ on the same set of axes, for $-4 \leq x \leq 4$, using 1 cm to represent 1 unit on both axes.

(ii) Read off the solution to the simultaneous equations
$x + y - 2 = 0$
$4y - x = 4$

Solution

(i) For each line choose three values of x and calculate the corresponding values of y. Then plot the lines and read off the coordinates of the point of intersection.

94

$x + y - 2 = 0$

x	−2	0	2
y	4	2	0

$4y - x = 4$

x	−4	0	4
y	0	1	2

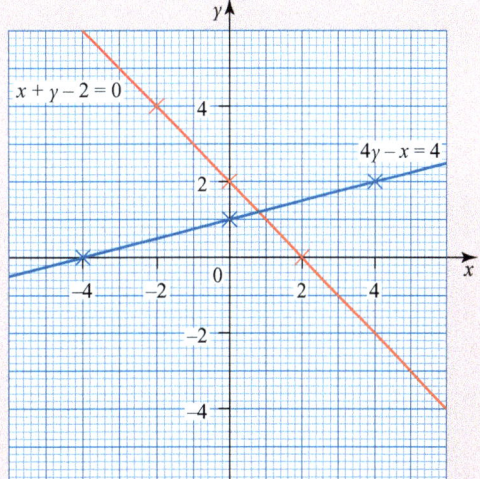

Figure 5.10

(11) The point of intersection is (0.8, 1.2), so the solution to the simultaneous equations is

$x = 0.8, y = 1.2$

Discussion points

→ Why should you plot three points for each line?

→ Two lines may not intersect. When is this the case?

Exercise 5C

You will need graph paper for this exercise.

① Solve these pairs of simultaneous equations by plotting their graphs. In each case you are given a suitable range of values of x.

(i) $x = 3y + 1$ $y = x - 1$ $0 \leq x \leq 3$

(ii) $3x + 2y = 5$ $x + y = 3$ $-2 \leq x \leq 2$

② Solve these pairs of simultaneous equations by plotting their graphs. In each case you are given a suitable range of values of x.

(i) $y = 2x - 4$ $3x + 4y = 17$ $0 \leq x \leq 6$

(ii) $6x + y = 1$ $4x - y = 4$ $0 \leq x \leq 2$

③ (i) Plot the lines $x = 4$, $y = x + 4$ and $4x + 3y = 12$ on the same axes for $-1 \leq x \leq 5$.

(ii) State the coordinates of the three points of intersection, and for each point give the pair of simultaneous equations that are satisfied there.

(iii) Work out the area of the triangle enclosed by the three lines.

④ (i) Using the same scale for both axes, plot the lines $2y + x = 4$ and $2y + x = 10$ on the same axes for $0 \leq x \leq 6$, and say what you notice about them. Why is this the case?

(ii) Add the line $y = 2x$ to your graph. What do you notice now? Can you justify what you see?

(iii) State the coordinates of the two points of intersection, and for each point give the pair of simultaneous equations that are satisfied there.

The intersection of two lines

5. A triangle has vertices A(0, 3), B(3, 6) and C(3, 0).
   (i) Work out the lengths of the sides of the triangle ABC.
   (ii) Work out the equations of the sides of the triangle ABC.
   (iii) Describe the triangle ABC.

6. A(1, 2), B(2, 5), C(5, 4) and D(4, 1) are the vertices of a quadrilateral ABCD.
   (i) Work out the gradients of the sides of the quadrilateral and state two pieces of information that this gives you.
   (ii) Work out the lengths of AB and BC.
   (iii) What type of quadrilateral is ABCD?

7. Alpha and Beta are two rival taxi firms which have the following price structures:

   Alpha: A fixed charge of £2 plus 60p per mile.

   Beta: A fixed charge of £3 plus 40p per mile.

   (i) On the same axes sketch the graph of price (vertical axis) against distance travelled (horizontal axis) for each firm.
   (ii) Write down the equation of each line.
   (iii) Which firm would you use for a distance of 7 miles?
   (iv) For what distance do both firms charge the same?

8. When the market price £p of an article varies, so does the number demanded, D and the number supplied, S.

   In one case $D = 15 + 0.5p$ and $S = p - 10$.

   (i) Sketch both lines on the same graph with D and S both on the vertical axis.

   The equilibrium position for the market is when the supply and the demand are equal.

   (ii) Work out the equilibrium price and the number bought and sold in equilibrium.

LEARNING OUTCOMES

Now you have finished this chapter, you should be able to
- identify parallel and perpendicular lines
- work out the equation of a line given the gradient of a line parallel or perpendicular to it and at least one point that it passes through
- calculate the distance between two points
- work out the coordinates of the midpoint of a line joining two points
- recognise different forms for the equation of a straight line
- find the point of intersection of two lines.

KEY POINTS

1 Two lines are parallel when their gradients are equal.
2 Two lines are perpendicular when the product of their gradients is −1.
3 When the points A and B have coordinates (x_1, y_1) and (x_2, y_2) respectively then
   - distance AB = $\sqrt{(x_2 - x_1)^2 + (y_2 - y_1)^2}$
   - midpoint of AB is $\left(\dfrac{x_1 + x_2}{2}, \dfrac{y_1 + y_2}{2}\right)$.
4 The coordinates of the point of intersection of two lines are found by solving their equations simultaneously.

6 Geometry I

> The difficulty lies, not in the new ideas, but in escaping the old ones, which ramify, for those brought up as most of us have been, into every corner of our minds.
>
> John Maynard Keynes

Prior knowledge

Much of the work in this chapter will already have been covered in your GCSE studies.

Knowledge of geometry topics will be needed within many sections of the specification.

This section provides a summary of the main facts that are required.

1 Pythagoras' theorem

$c^2 = a^2 + b^2$

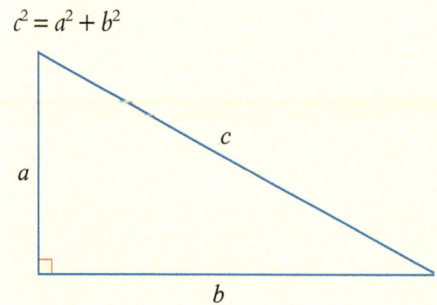

Figure 6.1

ACTIVITY 6.1

Write down the square of all the integers from 1 to 25 inclusive.
Check that $5^2 = 3^2 + 4^2$.
Write down as many other examples of $c^2 = a^2 + b^2$ as you can find.
How is each set of a, b and c linked to a right-angled triangle?

Pythagorean triples

The following are all Pythagorean triples as each set of three numbers satisfies $c^2 = a^2 + b^2$.

3, 4, 5   5, 12, 13   8, 15, 17   7, 24, 25

Using similar triangles, any multiple or fraction of each set will also be a Pythagorean triple.

For example,   9, 12, 15   2.5, 6, 6.5   16, 30, 34   1.4, 4.8, 5

2 Trigonometry in two dimensions

You have met definitions of the three trigonometric functions, sin, cos and tan, using the sides of a right-angled triangle.

sin is an abbreviation of sine, cos of cosine and tan of tangent.

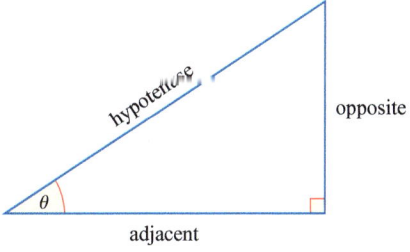

Figure 6.2

Discussion point

→ Do these definitions work for angles of any size?

In Figure 6.2

$$\sin\theta = \frac{\text{opposite}}{\text{hypotenuse}} \qquad \cos\theta = \frac{\text{adjacent}}{\text{hypotenuse}} \qquad \tan\theta = \frac{\text{opposite}}{\text{adjacent}}$$

ACTIVITY 6.2

(i) Using only a pencil, ruler and protractor, estimate $\sin 62°$.
(ii) Use your calculator to check your percentage error.
(iii) Suggest a way of reducing the percentage error when using this method.

Example 6.1

Work out the length of the side marked a in the triangle in Figure 6.3.

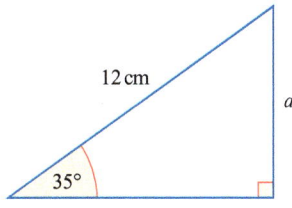

Figure 6.3

Trigonometry in two dimensions

Solution

Side a is *opposite* the angle of 35°, and the *hypotenuse* is 12 cm, so we use sin 35°.

$$\sin 35° = \frac{\text{opposite}}{\text{hypotenuse}}$$

$$= \frac{a}{12}$$

$$\Rightarrow \quad a = 12 \sin 35°$$

$$\Rightarrow \quad a = 6.9 \text{ cm (1 d.p.)}$$

Example 6.2

RWC

The diagram represents a ladder leaning against a wall.

Work out the length of the ladder.

Give your answer to 3 significant figures.

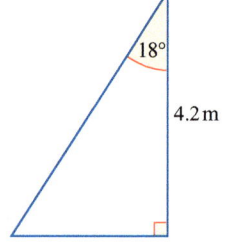

Figure 6.4

Solution

The side of length 4.2 m is *adjacent* to the angle of 18°, and we want the *hypotenuse* so use cos 18°.

$$\cos 18° = \frac{\text{adjacent}}{\text{hypotenuse}}$$

$$= \frac{4.2}{\text{hypotenuse}}$$

$$\text{hypotenuse} = \frac{4.2}{\cos 18°}$$

$$= 4.42 \text{ m (3 s.f.)}$$

Example 6.3

Work out the size of the angle marked θ in the triangle in Figure 6.5.

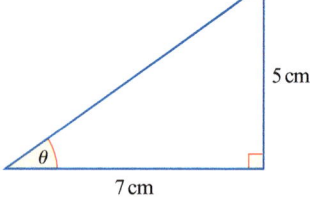

Figure 6.5

Discussion point

→ The full calculator value for $\frac{5}{7}$ has been used to work out the value of θ. What is the least number of decimal places that you could use to give the same value for the angle (to 1 d.p.) in this example?

Solution

The sides whose lengths are known are those *opposite* and *adjacent* to θ so we use tan θ.

$$\tan \theta = \frac{\text{opposite}}{\text{adjacent}} = \frac{5}{7}$$

$$\Rightarrow \quad \theta = 35.5° \text{ (1 d.p.)}$$

Example 6.4

A bird flies straight from the top of a 15 m tall tree, at an angle of depression of 27°, to catch a worm on the ground.

(i) How far does the bird fly?

(ii) How far was the worm from the bottom of the tree?

Solution

First draw a sketch, labelling the information given and using letters to mark what you want to find.

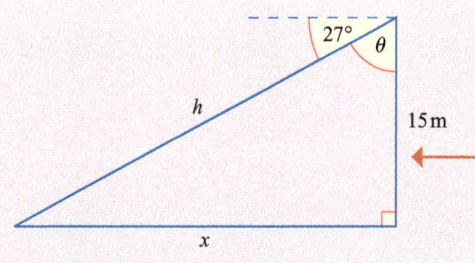

Figure 6.6

> **Note**
> Questions involving right-angled triangles will often entail applying trigonometry in a context. Examples and exercises include some questions without a context to provide practice of the skills needed in applications questions.

> Remember, *angles of depression* are measured <u>down</u> from the horizontal and angles of elevation are measured <u>up</u> from the horizontal.

(i) $\theta + 27° = 90°$

$\Rightarrow \quad \theta = 63°$

$\cos 63° = \dfrac{15}{h}$

$\Rightarrow \quad h = \dfrac{15}{\cos 63°} = 33.04033897$

❗ Make sure that you record the full calculator value of h for future use.

The bird flies 33 m.

(ii) Using Pythagoras' theorem

$h^2 = x^2 + 15^2$

$\Rightarrow \quad x^2 = 33.04033897^2 - 15^2 = 866.663999$

$\Rightarrow \quad x = 29.43915758$

The worm is 29.4 m from the bottom of the tree.

> **Discussion point**
> → If you used trigonometry for part (ii) of this question, which would be the best function to use? Why?

Historical note

The word for trigonometry is derived from three Greek words.

Tria: *three*   gonia: *angle*   metron: *measure*

(τρια)   (γονια)   (μετρον)

This shows how trigonometry developed from studying angles, often in connection with astronomy, although the subject was probably discovered independently by a number of people. Hipparchus (150 BC) is believed to have produced the first trigonometric tables which gave lengths of chords of a circle of unit radius. His work was further developed by Ptolemy in AD 100.

Trigonometry in two dimensions

Exercise 6A

① Work out the length marked x in each of these triangles. Give your answers correct to 1 decimal place.

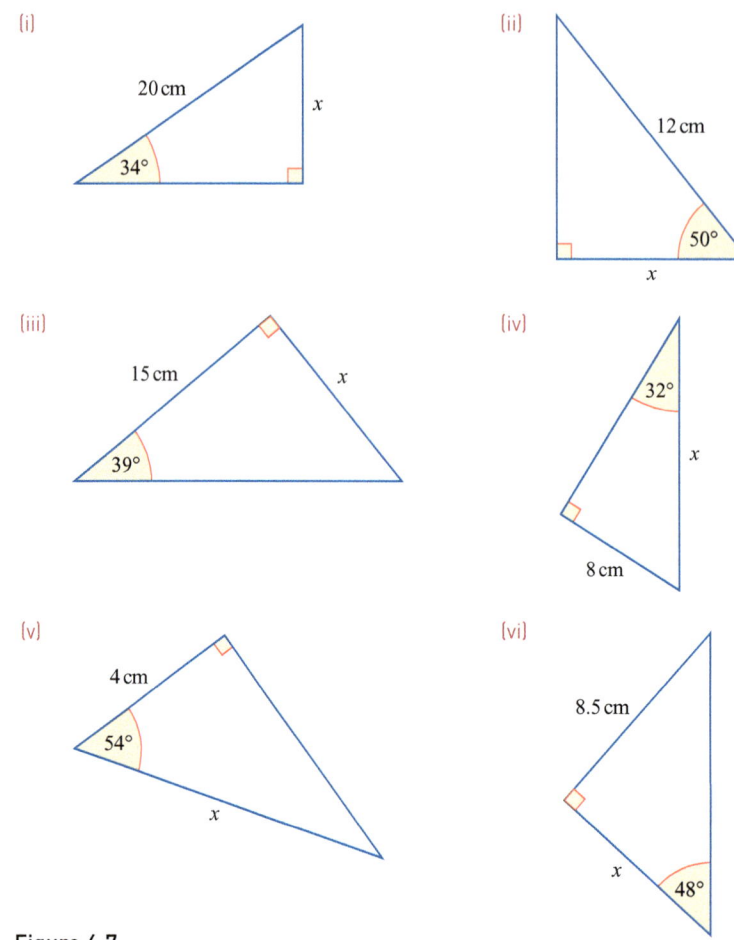

Figure 6.7

② Work out the size of the angle marked θ in each of these triangles. Give your answers correct to 1 decimal place.

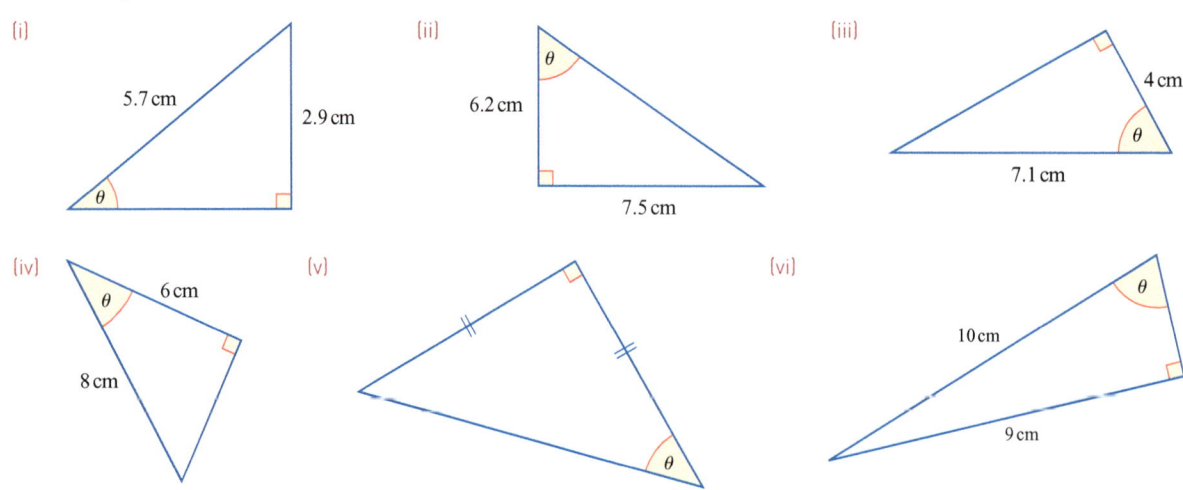

Figure 6.8

③ In an isosceles triangle, the line of symmetry bisects the base of the triangle. Use this fact to work out the angle θ and the lengths x and y in these diagrams.

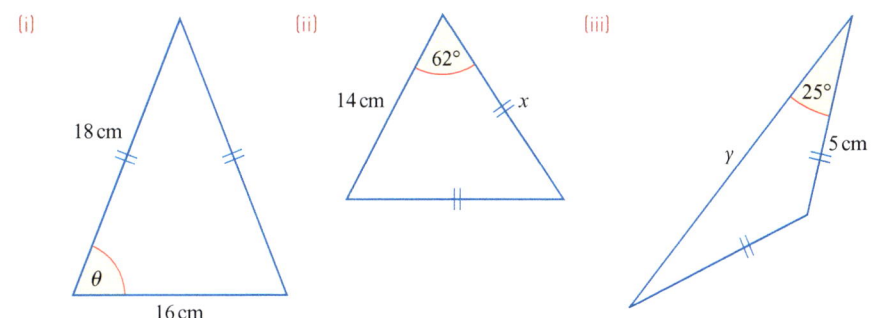

Figure 6.9

RWC ④ A ladder 5 m long rests against a wall. The foot of the ladder makes an angle of 65° with the ground.

How far up the wall does the ladder reach?

RWC ⑤ From the top of a vertical cliff 30 m high, the angle of depression of a boat at sea is 21°.

How far is the boat from the bottom of the cliff?

RWC ⑥ From a point 120 m from the base of an office block, the angle of elevation of the top of the block is 67°.

How tall is the block?

⑦ A rectangle has sides of length 12 cm and 8 cm.

What angle does the diagonal make with the longest side?

RWC ⑧ The diagram shows the positions of three airports:

E (East Midlands), M (Manchester) and L (Leeds).

The distance from M to L is 65 km on a bearing of 060°.

Angle LME = 90° and ME = 100 km.

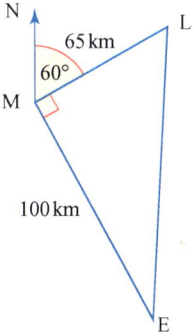

Figure 6.10

(i) Calculate, correct to 3 significant figures, the distance LE.

(ii) Calculate, correct to the nearest degree, the size of angle MEL.

(iii) An aircraft leaves M at 10.45 am and flies direct to E, arriving at 11.03 am.

Calculate, correct to 3 significant figures, the average speed of the aircraft in kilometres per hour.

Trigonometry in two dimensions

Angles of 45°, 30° and 60°

The sine, cosine and tangent of these angles have exact values.

When working without a calculator, the exact values should be known or derived.

Consider an isosceles right-angled triangle with AB = BC = 1 unit.

Discussion point
→ What would the results be if you used AB = BC = 2 units?

Using Pythagoras' theorem

$$AC^2 = 1^2 + 1^2$$
$$AC = \sqrt{2}$$

$$\sin 45° = \frac{\text{opp}}{\text{hyp}}$$

$$\sin 45° = \frac{1}{\sqrt{2}} \qquad \cos 45° = \frac{1}{\sqrt{2}} \qquad \tan 45° = 1$$

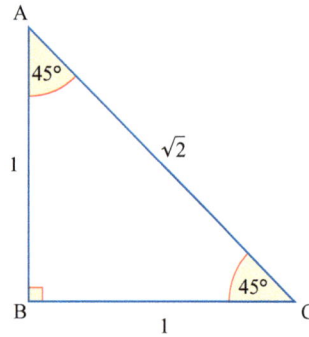

Figure 6.11

Consider an equilateral triangle of side length 2 units (Figure 6.12(a)).

By adding an angle bisector we get two congruent triangles (Figure 6.12(b)).

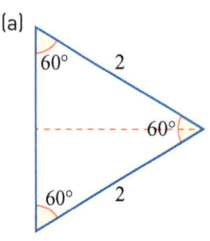
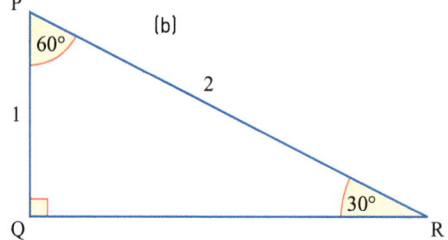

Figure 6.12

Using Pythagoras' theorem

$$QR^2 = 2^2 - 1^2$$
$$QR = \sqrt{3}$$

Using the trig ratios this gives us

$$\sin 30° = \frac{1}{2} \qquad \cos 30° = \frac{\sqrt{3}}{2} \qquad \tan 30° = \frac{1}{\sqrt{3}}$$

$$\sin 60° = \frac{\sqrt{3}}{2} \qquad \cos 60° = \frac{1}{2} \qquad \tan 60° = \sqrt{3}$$

Example 6.5

Do not use a calculator for this question.

Work out the exact value of y.

Give your answer in the form $p + q\sqrt{3}$ where p and q are integers.

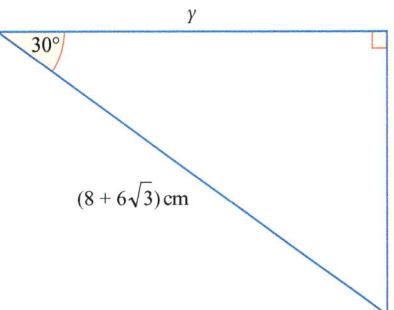

Figure 6.13

Solution

$$\cos 30° = \frac{y}{8 + 6\sqrt{3}}$$

$$\Rightarrow \quad \frac{\sqrt{3}}{2} = \frac{y}{8 + 6\sqrt{3}}$$

$$\Rightarrow \quad \frac{\sqrt{3}}{2} \times (8 + 6\sqrt{3}) = y$$

$$\Rightarrow \quad 4\sqrt{3} + 9 = y$$

$$y = 9 + 4\sqrt{3}$$

Exercise 6B

Use of a calculator is not allowed.

① Work out the exact value of x in each of the following.

Give your answers in their simplest form.

(i)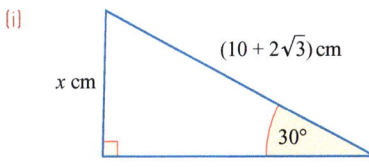

(ii)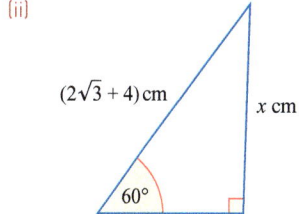

(iii)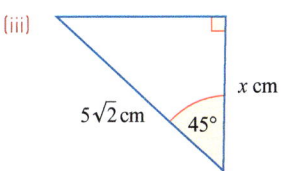

(iv)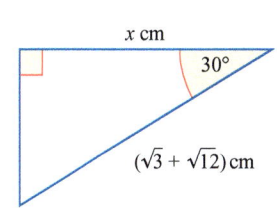

Figure 6.14

Trigonometry in two dimensions

② Look at Figure 6.15. Show that y is an integer.

③ Look at Figure 6.16. Show that p is an integer.

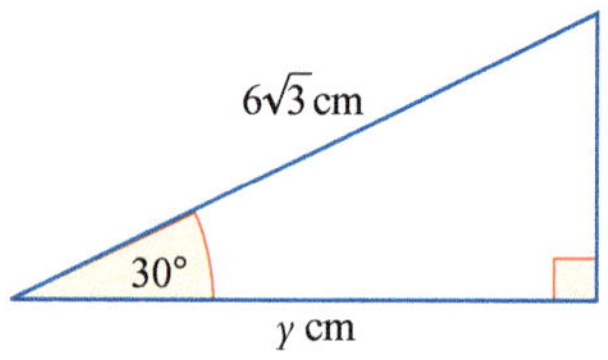

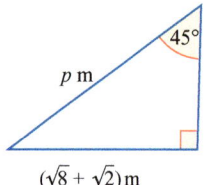

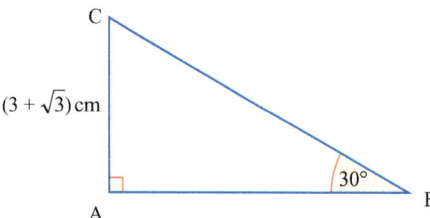

Figure 6.15

Figure 6.16

④ Look at Figure 6.17.

Figure 6.17

Work out the area of triangle ABC.

Give your answer in the form $p + q\sqrt{3}$ where p and q are integers.

⑤ Look at Figure 6.18. Work out the exact value of CD.

Give your answer in the form $k\sqrt{6}$ where k is an integer.

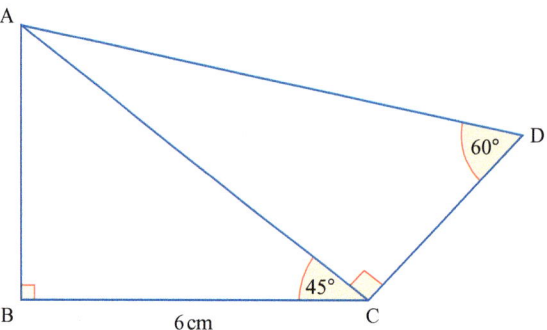

Figure 6.18

RWC ⑥ A ski-lift is spanning a valley in the Alps, rising from a height of 2039 m to a height of 2364 m over a horizontal distance of 325 m. What is the angle of elevation of the ski lift?

RWC ⑦ The centrepiece of a show garden has been designed as a square of side 3 metres surrounded by four equilateral triangles as shown in Figure 6.19.

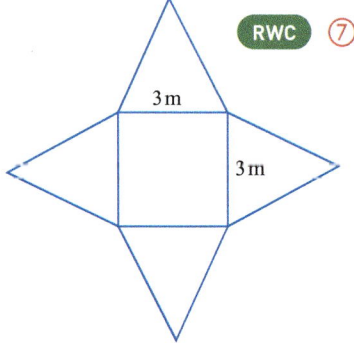

Figure 6.19

(i) The centre square is to be planted with small shrubs which each require a square area of side 30 cm. How many shrubs are required?

(ii) The triangular areas are to be planted with bedding plants, each requiring an area of approximately 100 cm². Approximately how many bedding plants will be required?

(iii) The bedding plants are sold in boxes of 12 and the head gardener decides to order 5% extra plants to allow for ones which might not be up to standard. How many boxes does she need to order?

⑧ Figure 6.20 shows a vertical building standing on horizontal ground. The points A, B and C are in a straight line on horizontal ground and AC = 30 m. The point T is at the top of the building and CT is vertical. The angles of elevation of T from A and B are 30° and 60° respectively.

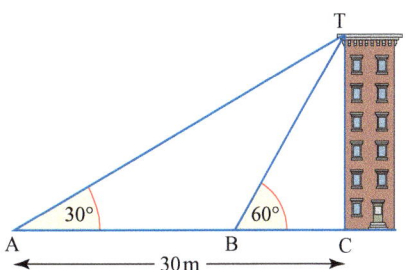

Figure 6.20

(i) Calculate the exact value of the height CT of the building.

(ii) Work out the distances BC and AB.

3 Trigonometric functions for angles of any size

By convention, angles are measured anticlockwise from the positive x-axis (Figure 6.21). Anticlockwise is taken to be positive and clockwise to be negative.

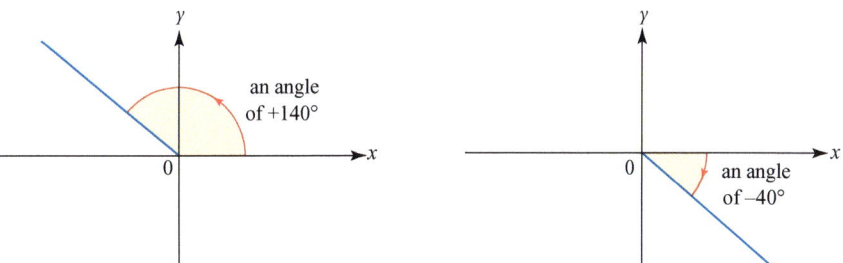

Figure 6.21

The only exception is for compass bearings, which are measured clockwise from the north.

Definitions of the trigonometric functions, sin, cos and tan

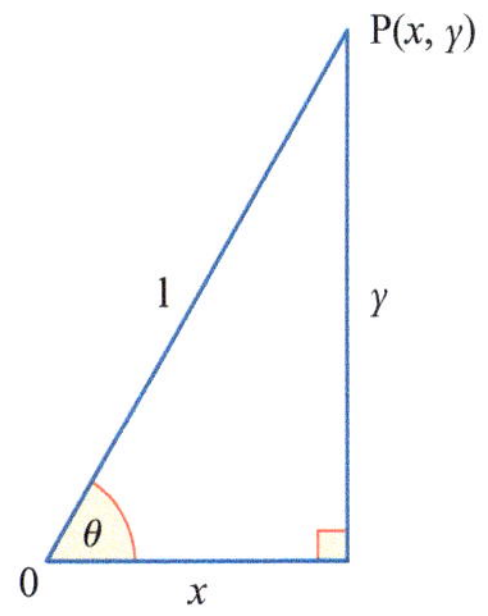

First look at the right-angled triangle in Figure 6.22 which has a hypotenuse of unit length.

Sin, cos and tan are defined as the following ratios.

$$\sin\theta = \frac{y}{1} = y \qquad \cos\theta = \frac{x}{1} = x \qquad \tan\theta = \frac{y}{x}$$

We can extend these definitions to angles beyond 90°.

Figure 6.22

The sine and cosine graphs

Imagine the angle θ situated at the origin, as in Figure 6.23, and allow θ to take any value. The vertex marked P has coordinates $(\cos\theta, \sin\theta)$ and can now be anywhere on the unit circle.

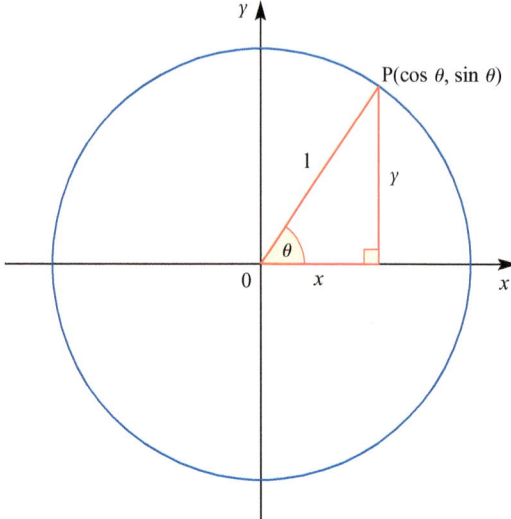

Figure 6.23

You can now see that these definitions can be applied to *any* angle θ, whether it is positive or negative, and whether it is less than or greater than 90°.

$$\sin\theta = y \quad \cos\theta = x \quad \tan\theta = \frac{y}{x}$$

For some angles, x or y (or both) will take a negative value, so the signs of $\sin\theta$, $\cos\theta$ and $\tan\theta$ will vary accordingly.

4 The sine and cosine graphs

Look at Figure 6.24. There is a unit circle, and angles have been drawn at intervals of 30°. The resulting y-coordinates are plotted relative to the axes on the right. They have been joined with a continuous curve to give the graph of $\sin\theta$ for $0 \leq \theta \leq 360°$.

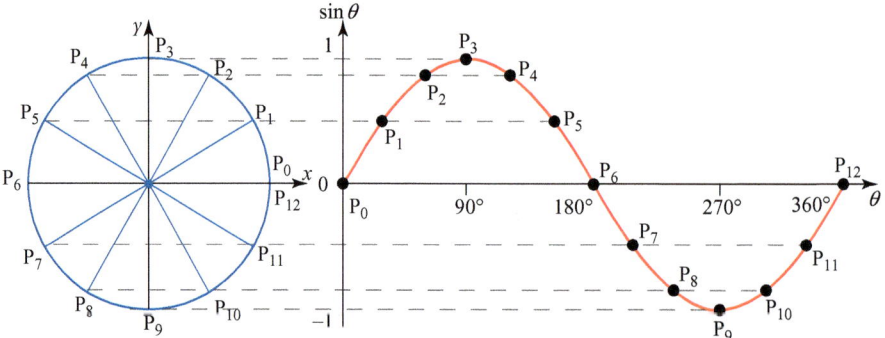

Figure 6.24

Continuing this process for angles 390°, 420°, ... and angles −30°, −60°, ... you get the graph of $y = \sin\theta$, as shown in Figure 6.25.

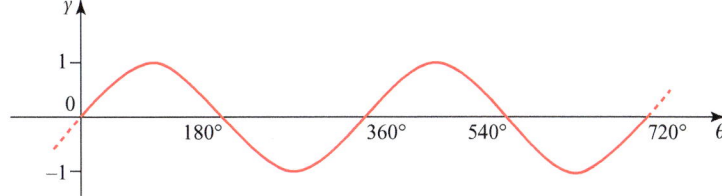

Figure 6.25

Since the curve repeats itself every 360°, the sine function is described as *periodic* with *period* 360°.

In a similar way you can transfer the *x*-coordinates onto a set of axes to obtain the cosine graph. This is most easily illustrated if you first rotate the circle through 90° anticlockwise.

Figure 6.26 shows this new orientation, together with the resulting graph.

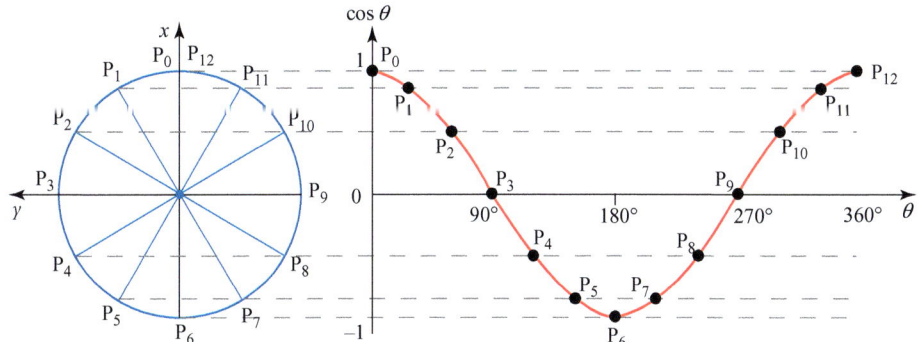

Figure 6.26

For angles beyond this interval the cosine graph repeats itself periodically, with a period of 360°.

Notice that the graphs of $\sin\theta$ and $\cos\theta$ have exactly the same shape. The cosine graph can be obtained by translating the sine graph 90° to the left, as shown in Figure 6.27.

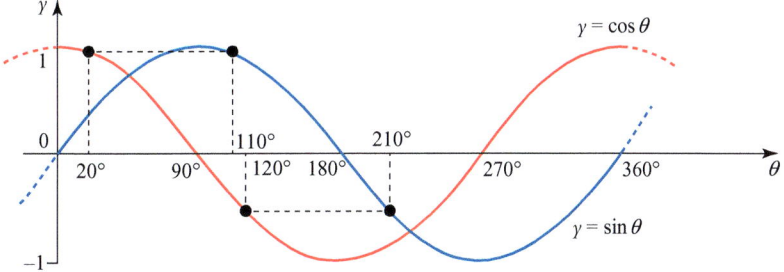

Figure 6.27

Solution of trigonometric equations

5 The tangent graph

The value of $\tan\theta$ can be worked out from the definition $\tan\theta = \dfrac{y}{x}$ or by using $\tan\theta = \dfrac{\sin\theta}{\cos\theta}$.

> **Discussion points**
> → The function $\tan\theta$ is undefined for $\theta = 90°$. What does *undefined* mean?
> → How can you tell that $\tan 90°$ is undefined?
> → For which **other** values of θ is $\tan\theta$ undefined?

The graph of $\tan\theta$ is shown in Figure 6.28. The dotted lines $\theta = \pm 90°$ and $\theta = 270°$ are *asymptotes*; they are not actually part of the curve.

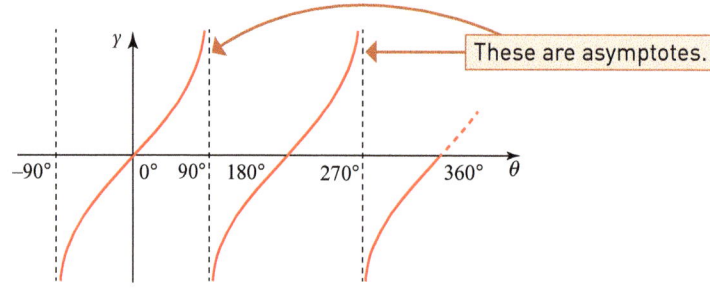

> **Discussion points**
> How would you describe an asymptote to a friend?

Figure 6.28

> **Note**
> It is important to learn the graphs of $y = \sin\theta$, $y = \cos\theta$ and $y = \tan\theta$.

> **Discussion points**
> → The graph of $\tan\theta$ is periodic, like those for $\sin\theta$ and $\cos\theta$. What is the period of this graph?
> → Show how the part of the curve for $0° \leqslant \theta \leqslant 90°$ can be used to generate the rest of the curve using rotations and translations.

6 Solution of trigonometric equations

Suppose that you want to solve the equation

$$\sin\theta = 0.5$$

You start by pressing the calculator keys

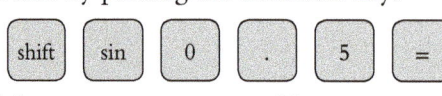

> **Note**
> The $\sin^{-1}$ key may also be labelled invsin or arcsin.

and the answer comes up as 30.

If your calculator does not give the answer 30 then it might be in the wrong angle setting. Check for a D (or DEG) at the top of the screen. If not, then select DEG whilst in SETUP mode.

However, look at the graph of $y = \sin\theta$ (Figure 6.29). You can see that there are other roots as well.

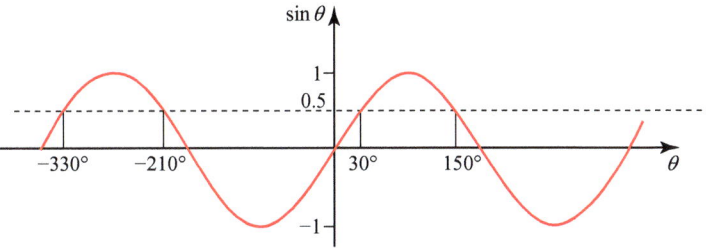

> **Discussion point**
> → How many roots does the equation have?

Figure 6.29

The root 30° is called the *principal value*.

Other roots can be found by looking at the graph. The roots for $\sin\theta = 0.5$ are seen to be

$$\theta = \ldots, -330°, -210°, 30°, 150°, \ldots$$

As the graph is periodic, then the roots repeat every 360°.

> **Note**
>
> A calculator always gives the principal value of the solution. These values are in the range
>
> $0° \leq \theta \leq 180°$ (cos) $-90° \leq \theta \leq 90°$ (sin) $-90° < \theta < 90°$ (tan).

Example 6.6

Work out the values of θ in the interval $0° \leq \theta \leq 360°$ for which $\cos\theta = 0.4$

Solution

$\cos\theta = 0.4 \Rightarrow \theta = 66.4°$ (principal value).

Figure 6.30 shows the graph of $y = \cos\theta$.

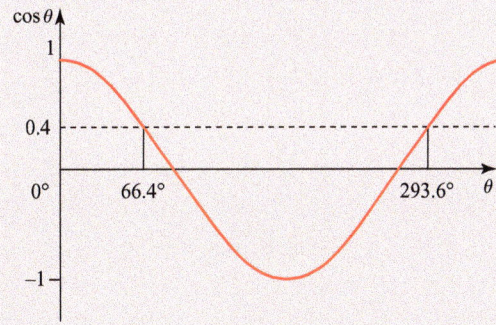

Figure 6.30

The values of θ for which $\cos\theta = 0.4$ are 66.4°, 293.6°.

> **Discussion points**
> → How do we get 293.6° from 66.4°?
> → Is there a general rule for finding a second angle between $0° \leq \theta \leq 360°$?

Solution of trigonometric equations

Example 6.7

Work out the values of x in the interval $-360° \leq x \leq 360°$ for which $6 + 2\tan x = 0$.

Solution

$$6 + 2\tan x = 0$$
$$\Rightarrow \quad 2\tan x = -6$$
$$\Rightarrow \quad \tan x = -3$$
$$\Rightarrow \quad x = -71.6° \text{ (principal value)}$$

Figure 6.31 shows the graph of $y = \tan x$.

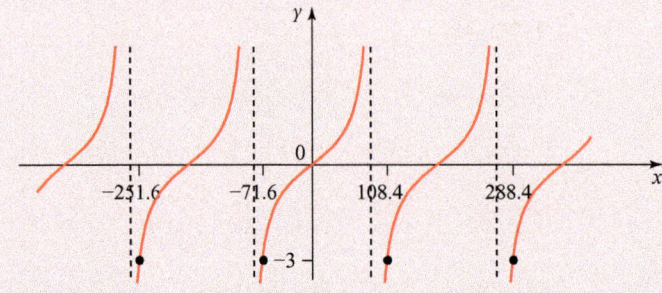

Figure 6.31

The values of x for which $\tan x = -3$ are
$-251.6°, -71.6°, 108.4°, 288.4°$.

Short method for solving trigonometric equations for any angle:

Step 1: Use $\sin^{-1}$, $\cos^{-1}$ or $\tan^{-1}$ to find the principal value, θ.

Step 2: Work out a second angle using one of the following

$$\sin^{-1}: \quad 180° - \theta$$
$$\cos^{-1}: \quad 360° - \theta$$
$$\tan^{-1}: \quad \theta + 180°$$

Step 3: Add $360°$ to both values, until the upper limit is reached.

Step 4: Subtract $360°$ from both values, until the lower limit is reached.

Example 6.8

Solve $10\sin\theta + 3 = 0$ for $-360° \leq \theta \leq 720°$.

Solution

$$10\sin\theta + 3 = 0$$
$$\Rightarrow \quad \sin\theta = -0.3$$

Step 1: Principal value $= \sin^{-1}(-0.3)$
$$= -17.5°$$

Step 2: The second angle is $180° + 17.5° = 197.5°$.

Step 3: $-17.5° + 360° = 342.5°$ and $342.5° + 360° = 702.5°$
$197.5° + 360° = 557.5°$

Step 4: $-17.5° - 360° = -377.5°$ (too low)
$197.5° - 360° = -162.5°$

∴ $\theta = -162.5°, -17.5°, 197.5°, 342.5°, 557.5°$ or $702.5°$.

ACTIVITY 6.3

> Make sure the graph plotter is set to 'degrees'.

- Use a graph plotter to plot the graph of $y = \sin(x + 10)$.
- How does the graph of $y = \sin(x + 10)$ compare with the graph of $y = \sin x$?
- Then do the same with the graph of $y = \sin(x + 20)$.
- Is it possible to find a graph of the form $y = \sin(x + c)$ which is the same as the graph of $y = \cos x$?
- Write down an identity in the form $\cos x \equiv \sin(x + c)$ where c is a number to be found.
- Go through the same process with the cosine graph to find a similar identity of the form $\sin x \equiv \cos(x + d)$.

> **Note**
> Knowledge of these graphs and the corresponding identities will not be examined in this specification. However, students who go on to study Mathematics at A-Level will meet them again.

Exercise 6C

Give answers to 1 decimal place where necessary.

① Solve the following equations for $0° \leq \theta \leq 360°$.

(i) $\cos \theta = 0.5$ (ii) $\tan \theta = 1$ (iii) $\sin \theta = \dfrac{\sqrt{3}}{2}$

(iv) $\sin \theta = -0.5$ (v) $\cos \theta = 0$ (vi) $\tan \theta = -5$

(vii) $\tan \theta = 0$ (viii) $\cos \theta = -0.54$ (ix) $\sin \theta = 1$

② Solve the following equations for $-180° \leq \theta \leq 180°$.

(i) $3\cos \theta = 2$ (ii) $7\sin \theta = 5$ (iii) $3\tan \theta = 8$

(iv) $6\sin \theta + 5 = 0$ (v) $5\cos \theta + 2 = 0$ (vi) $5 - 9\tan \theta = 10$

SA ③ Solve the following equations for $0° \leq \theta \leq 360°$.

(i) $\sin^2 \theta = 0.75$ (ii) $\cos^2 \theta = 0.5$ (iii) $\tan^2 \theta = 1$

SA ④ (i) Factorise $2x^2 + x - 1$.

(ii) Hence solve $2x^2 + x - 1 = 0$.

(iii) Use your results to solve these equations for $-360° \leq \theta \leq 360°$.

(a) $2\sin^2 \theta + \sin \theta - 1 = 0$

(b) $2\cos^2 \theta + \cos \theta - 1 = 0$

(c) $2\tan^2 \theta + \tan \theta - 1 = 0$

> $\sin^2 \theta$ is alternative notation for $(\sin \theta)^2$.

SA ⑤ Solve the following equations for $-180° \leq x \leq 180°$.

(i) $\tan^2 x - 3\tan x = 0$ (ii) $1 - 2\sin^2 x = 0$

(iii) $3\cos^2 x + 2\cos x - 1 = 0$ (iv) $2\sin^2 x = \sin x + 1$

(6) Do not use a calculator in this question.

Solve the following equations for $-360° < x < 360°$.

(i) $\tan x = \sqrt{3}$
(ii) $2\sin x = 1$
(iii) $\sqrt{2}\cos x - 1 = 0$
(iv) $2\sin x = \sqrt{3}$
(v) $\tan^2 x - \tan x = 0$
(vi) $4\cos x = \sqrt{12}$

(7) Solve $(\cos\theta - 1)(\cos\theta + 2)(2\cos\theta - 1) = 0$ for $0° \leq \theta \leq 360°$.

(8) (i) Given that $f(x) = 2x^3 - x^2 - 3x - 1$, calculate $f\left(-\dfrac{1}{2}\right)$.

(ii) Hence solve $2\sin^3\theta - \sin^2\theta - 3\sin\theta - 1 = 0$ for $-180° \leq \theta \leq 180°$.

7 $a\sin kx$, $a\cos kx$ and $a\tan kx$

Transformations of $y = a\sin kx$, $y = a\cos kx$ and $y = a\tan kx$

$y = \sin x \mapsto y = a\sin x$
$y = \cos x \mapsto y = a\cos x$
$y = \tan x \mapsto y = a\tan x$

Each of these transformations is a vertical stretch with scale factor a.

A negative scale factor will also include a reflection in the x-axis.

$y = \sin x \mapsto y = \sin(kx)$
$y = \cos x \mapsto y = \cos(kx)$
$y = \tan x \mapsto y = \tan(kx)$

Each of these transformations is a horizontal stretch with scale factor $\dfrac{1}{k}$

Example 6.9

Sketch the graphs of $y = \sin x$ and $y = 2\sin x$ for $0° \leq x \leq 360°$ on the same pair of axes.

A vertical stretch of scale factor 2 away from the x-axis.

Solution

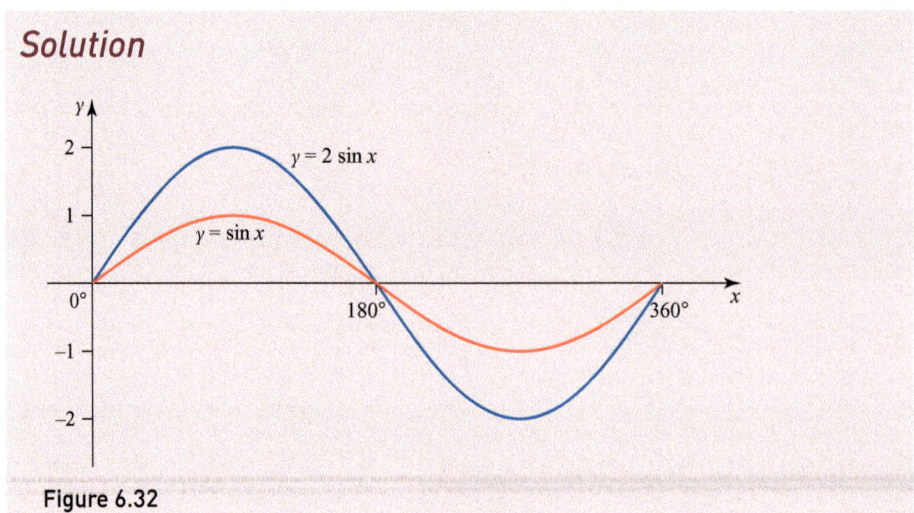

Figure 6.32

Example 6.10

Sketch the graphs of $y = \cos x$ and $y = \cos 3x$ for $0° \leq x \leq 360°$ on the same pair of axes.

A horizontal stretch of scale factor $\frac{1}{3}$.

Solution

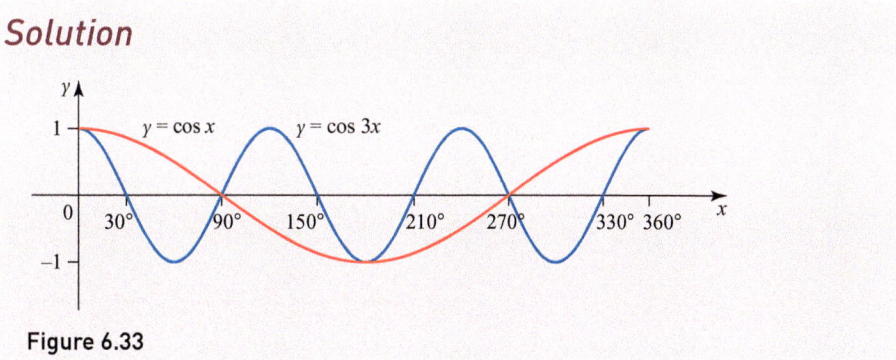

Figure 6.33

Example 6.11

A reflection in the x-axis *and* a vertical stretch (s.f. 2) *and* a horizontal stretch (s.f. 3).

Sketch the graphs of $y = \tan x$ and $y = -2\tan\left(\frac{1}{3}x\right)$ for $0° \leq x \leq 360°$ on the same pair of axes.

Solution

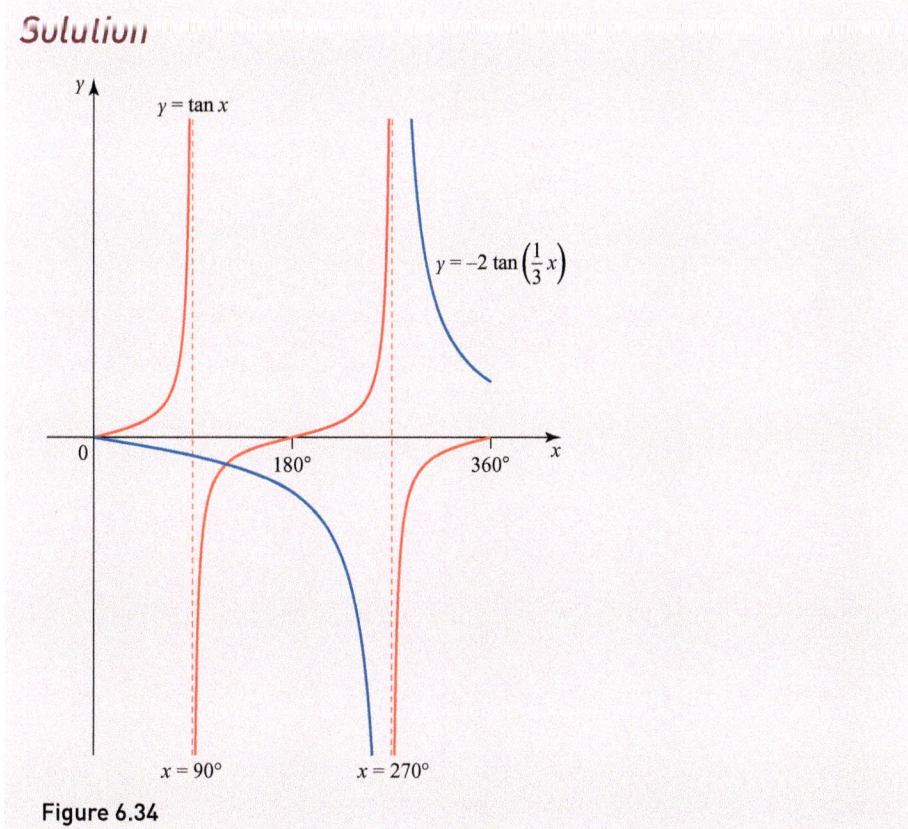

Figure 6.34

Trigonometric equations

If a trigonometric graph has already been drawn, then solving an equation can often be easier by considering the intersection points with the graph. The symmetrical and periodic properties of trigonometric functions can be used to find multiple solutions.

a sin kx, a cos kx and a tan kx

Example 6.12

(i) Sketch the graph of $y = 2\sin x$ for $0° \leqslant x \leqslant 360°$.

(ii) Find all the solutions of the equation $2\sin x = 1$ for $0° \leqslant x \leqslant 360°$.

(iii) Write down the maximum and minimum values of $y = 2\sin x$.

Solution

(i)

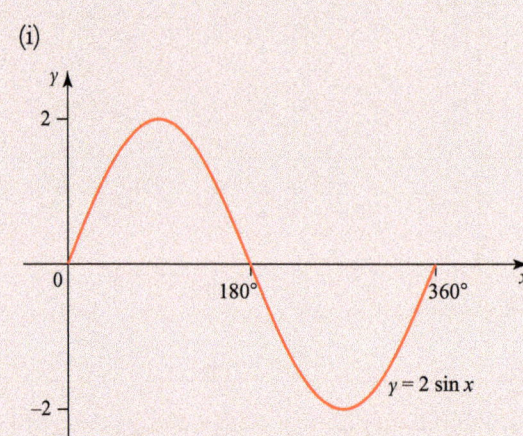

Figure 6.35

(ii) The first solution can be found by rearranging the equation and then using a calculator.

$$2\sin x = 1 \implies \sin x = \frac{1}{2}$$

$$\implies x = \sin^{-1}\left(\frac{1}{2}\right) = 30°$$

Then consider the intersection of the graphs $y = 2\sin x$ and $y = 1$.

The first intersection must be when $x = 30°$.

Using symmetry, the other intersection must be when $x = 180 - 30 = 150°$.

So the solutions are $x = 30°, 150°$.

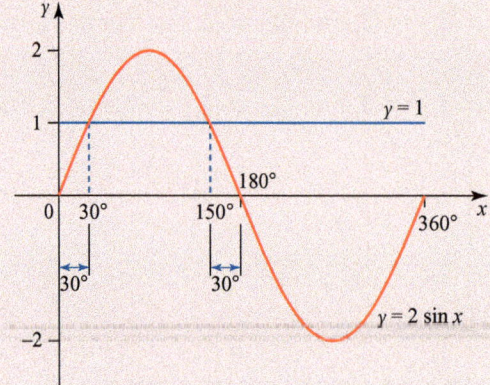

Figure 6.36

(iii) The maximum value is 2.

The minimum value is −2.

Example 6.13

(i) Sketch the graph of $y = 3\cos 2x$ for $0° \leq x \leq 360°$.

(ii) Find all solutions of the equation $3\cos 2x = -2$ for $0° \leq x \leq 360°$.

(iii) Write down the maximum and minimum values of $y = 3\cos 2x$.

Solution

(i)

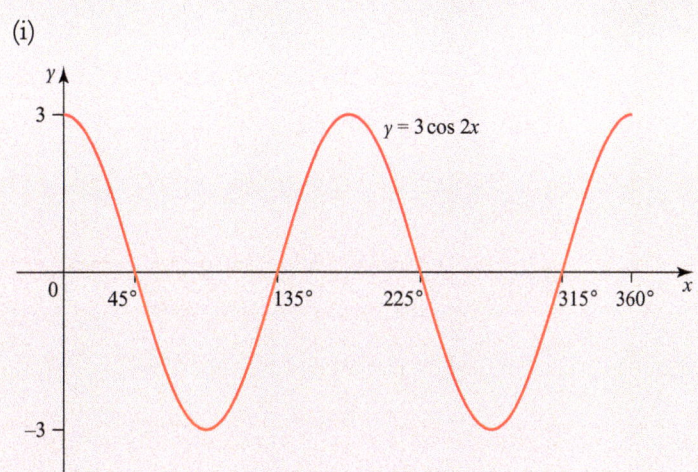

Figure 6.37

(ii) The first solution can be found by rearranging the equation and then using a calculator.

$$3\cos 2x = -2 \implies \cos 2x = -\tfrac{2}{3}$$
$$\implies 2x = \cos^{-1}\left(-\tfrac{2}{3}\right) = 131.8°$$
$$\implies x = 65.9°$$

Then consider the intersection of the graphs $y = 3\cos 2x$ and $y = -2$.

The first intersection must be when $x = 65.9°$.

Using symmetry, the next intersection must be when $x = 135 - 20.9 = 114.1°$.

The next intersections must be when

$x = 225 + 20.9 = 245.9°$ and $x = 315 - 20.9 = 294.1°$.

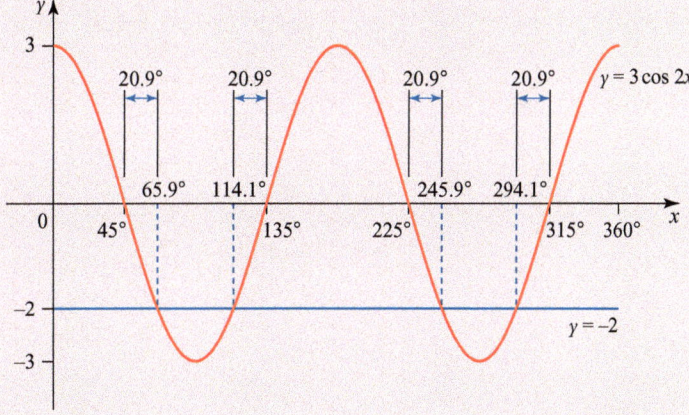

Figure 6.38

$a \sin kx$, $a \cos kx$ and $a \tan kx$

So the solutions are $x = 65.9°, 114.1°, 245.9°, 294.1°$ (1 d.p.).

(iii) The maximum value is 3.

The minimum value is -3.

Example 6.14

(i) Sketch the graph of $y = 4\sin x + 3$ for $0° \leqslant x \leqslant 360°$.

Include the coordinates of any intersections with the axes.

(iii) Write down the maximum and minimum values of $y = 4 \sin x + 3$.

Solution

(i) $4\sin x$ takes values between -4 and 4.

So $4\sin x - 3$ takes values between 1 and -7.

The y-intercept is $0 - 3 = -3$.

The graph intersects the x-axis when

$4\sin x - 3 = 0 \Rightarrow \sin x = \dfrac{3}{4}$

$\Rightarrow x = 48.6°$

The other x-intercept is when

$x = 180 - 48.6 = 131.4°$.

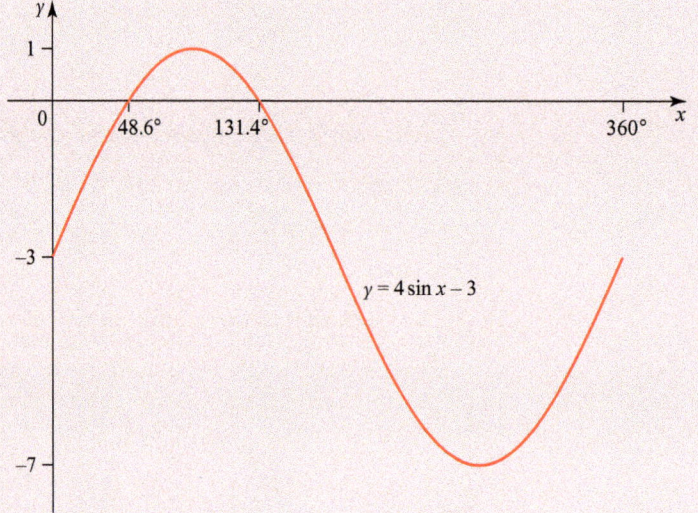

Figure 6.39

(ii) The maximum value is 1.

The minimum value is -7.

Exercise 6D

① Sketch the following pairs of graphs on the same axes.
Indicate the coordinates of any axis intercepts.
Include any asymptotes and their equations.
(i) $y = \sin x$ and $y = 3\sin x$
(ii) $y = \tan x$ and $y = \tan 2x$
(iii) $y = \cos x$ and $y = -2\cos x$

② Four of the following five equations are represented as graphs in Figure 6.40.

$y = 4\sin x$ $\quad y = 4\cos x$ $\quad y = \sin 2x$ $\quad y = \cos\left(\frac{1}{2}x\right)$ $\quad y = 2\sin x$

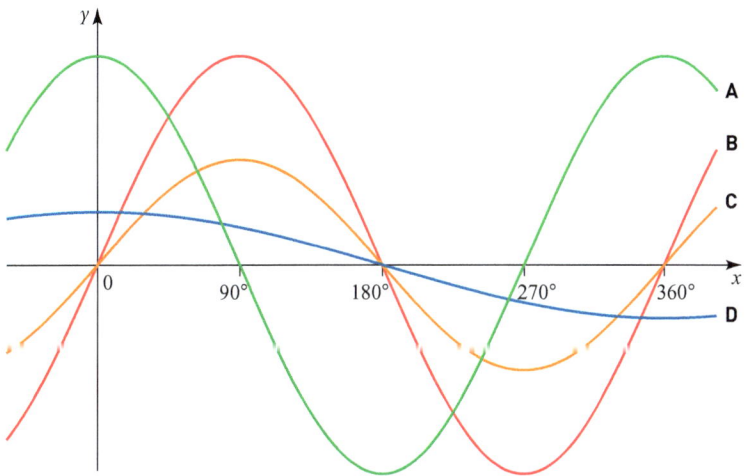

Figure 6.40

(i) Match each graph with its equation.
(ii) Sketch a graph of the equation which is not matched.

③ Write down the maximum and minimum values of each of these expressions.
(i) $\sin 4x$ (ii) $3\cos x$ (iii) $2\sin 5x$
(iv) $-6\cos 7x$ (v) $4\sin x + 9$ (vi) $-5\cos 6x + 1$

④ (i) Sketch the graph of $y = \sin 3x$ for $0° \leqslant x \leqslant 360°$.
(ii) Find all solutions of the equation $\sin 3x = 1$ for $0° \leqslant x \leqslant 360°$.

⑤ (i) Sketch the graph of $y = \cos 2x$ for $0° \leqslant x \leqslant 360°$.
(ii) Find all solutions of the equation $\cos 2x = 0.5$ for $0° \leqslant x \leqslant 360°$.

⑥ (i) Sketch the graph of $y = 2\cos 3x$ for $0° \leqslant x \leqslant 360°$.
(ii) Find all solutions of the equation $2\cos 3x = -1$ for $0° \leqslant x \leqslant 360°$.

⑦ (i) Sketch the graph of $y = 5\sin 2x + 1$ for $0° \leqslant x \leqslant 360°$.
Include the coordinates of any intersections with the axes.
(ii) Write down the maximum and minimum values of $y = 5\sin 2x + 1$.

⑧ (i) Sketch the graph of $y = -4\cos 3x - 3$ for $0° \leqslant x \leqslant 360°$.
Include the coordinates of any intersections with the axes.
(ii) Write down the maximum and minimum values of $y = -4\cos 3x - 3$.

Learning outcomes

FUTURE USES

Trigonometric functions are explored to a greater depth in A-Level Mathematics, including the use of various trig identities and the study of inverse trig functions. Trigonometry is used in many areas, including mechanics when resolving vectors such as forces and velocities.

It is also used extensively in A-Level Further Mathematics, to describe complex numbers (a combination of real and imaginary numbers), to describe transformations, and many other applications.

In A-Level Further Mathematics, you will also study the hyperbolic functions $\sinh x$, $\cosh x$ and $\tanh x$.

REAL-WORLD CONTEXT

Trigonometry has many real-world applications, including every aspect of engineering. It is also essential to architects and surveyors. Space exploration and the motion/positioning of satellites would not be possible without trigonometry. Mobile telephones, video games, and computers in general, make much use of this vital area of mathematics. In fact, ancient civilisations were aware of its usefulness, and made use of it to achieve amazing feats of construction, many of which are still standing to this day.

LEARNING OUTCOMES

Now you have finished this chapter, you should be able to
- use trigonometry in a right-angled triangle
  - to find an angle when you know any two sides
  - to find the other sides or angle when you know the length of one side and an angle
- use Pythagoras' theorem in two dimensions
  - in the form $a^2 + b^2 = c^2$
- solve practical problems in two dimensions (e.g. a ladder against a wall)
- use a calculator to find
  - the sin, cos or tan of any angle
  - an angle, given the sin, cos or tan ratio
- sketch and recognise the graphs of sin, cos or tan for any angle
- solve trigonometric equations
- stretch, reflect and translate trigonometric graphs.

KEY POINTS

1 In a right-angled triangle Pythagoras' theorem gives
$c^2 = a^2 + b^2$

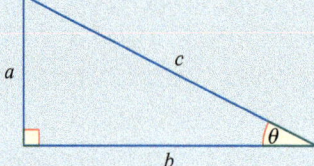

Figure 6.41

2 Using the triangle above gives the definitions:

$$\sin \theta = \frac{a}{c} \quad \cos \theta = \frac{b}{c} \quad \tan \theta = \frac{a}{b}$$

4 $\sin 45° = \dfrac{1}{\sqrt{2}}$ $\cos 45° = \dfrac{1}{\sqrt{2}}$ $\tan 45° = 1$

   $\sin 30° = \dfrac{1}{2}$ $\cos 30° = \dfrac{\sqrt{3}}{2}$ $\tan 30° = \dfrac{1}{\sqrt{3}}$

   $\sin 60° = \dfrac{\sqrt{3}}{2}$ $\cos 60° = \dfrac{1}{2}$ $\tan 60° = \sqrt{3}$

5 Trig graphs:

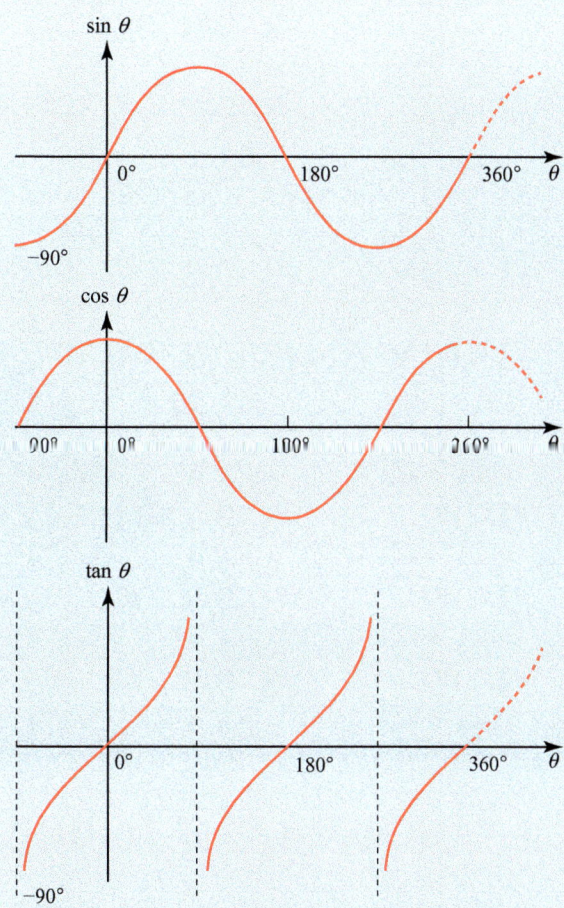

Figure 6.42

6 The graph of $y = a\sin(kx)$ is two stretches of the graph of $y = \sin(x)$: a vertical stretch scale factor a, and a horizontal stretch scale factor $\dfrac{1}{k}$

   and

   $y = a\sin(kx)$ has a maximum value of a and a minimum value of $-a$.

Geometry II

What we know is a drop; what we don't know is an ocean.

Isaac Newton

1 The area of a triangle

You are familiar with the use of capital letters to label the vertices of a triangle. In a similar way you can use lower case letters to label the sides.

a denotes the length of the side opposite angle A, b is the length of the side opposite angle B, and c is the length of the side opposite angle C.

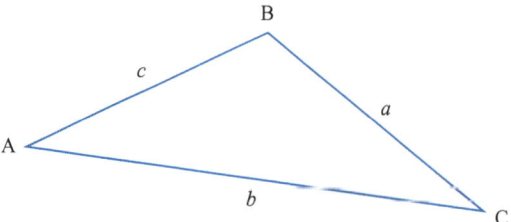

Figure 7.1

Using this notation, for any triangle ABC the area is given by the formula

$$\text{area} = \tfrac{1}{2} bc \sin A.$$

Proof

Figure 7.2 shows a triangle ABC. The perpendicular CD is the height h corresponding to AB as base.

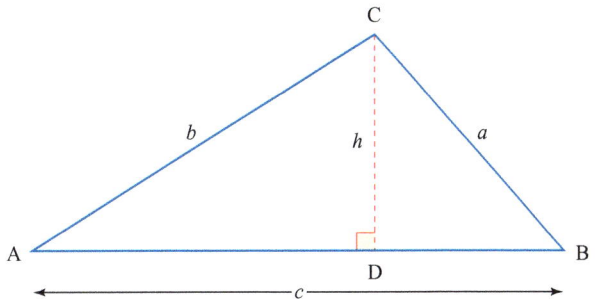

Figure 7.2

Using 'area of a triangle equals half its base times its height',

$$\text{area} = \tfrac{1}{2}ch. \qquad ①$$

In triangle ACD

$$\sin A = \frac{h}{b}$$

$$\Rightarrow \quad h = b \sin A.$$

Substituting in ① gives

$$\text{area} = \tfrac{1}{2}bc \sin A.$$

> **Note**
>
> Taking the other two points in turn as the top of the triangle gives equivalent results:
> $$\text{area} = \tfrac{1}{2}ca \sin B$$
> and
> $$\text{area} = \tfrac{1}{2}ab \sin C.$$
> The formula may be easier to remember as 'half the product of two sides times the sine of the angle between them'.

The area of a triangle

Example 7.1

Figure 7.3 shows a regular pentagon, PQRST, inscribed in a circle, centre C, radius 8 cm. Calculate the area of

(i) triangle CPQ

(ii) the pentagon.

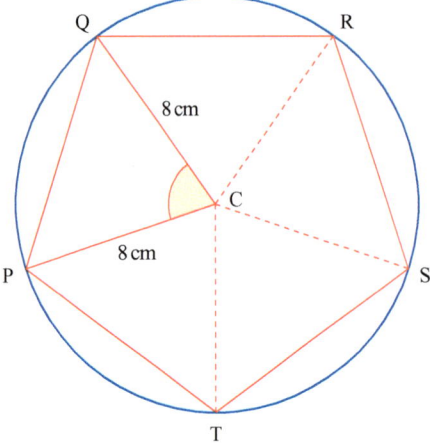

Figure 7.3

Solution

(i) angle PCQ = 360° ÷ 5
   = 72°

   area PCQ = $\frac{1}{2} \times 8 \times 8 \times \sin 72°$
   = 30.4338...
   = 30.4 cm² (1 d.p.)

(ii) area PQRST = 5 × 30.4338...
   = 152.169...
   = 152.2 cm² (1 d.p.)

Example 7.2

Figure 7.4 shows an isosceles triangle with an area of 24 cm² and one angle of 40°. Calculate the lengths of the two equal sides.

Solution

Let the equal sides be of length x cm.

Using   area = $\frac{1}{2} ab \sin C$

∴ $24 = \frac{1}{2} \times x \times x \times \sin 40°$

⇒ $x^2 = \dfrac{48}{\sin 40°}$

⇒ $x = 8.64$ cm   (3 s.f.)

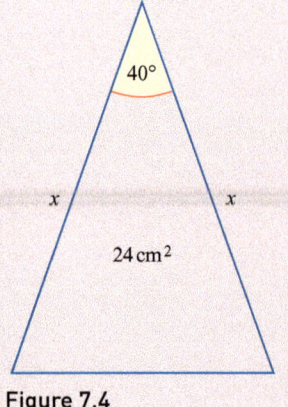

Figure 7.4

Exercise 7A

Where necessary leave answers approximated to 3 significant figures.

① Work out the area of each of the following triangles.

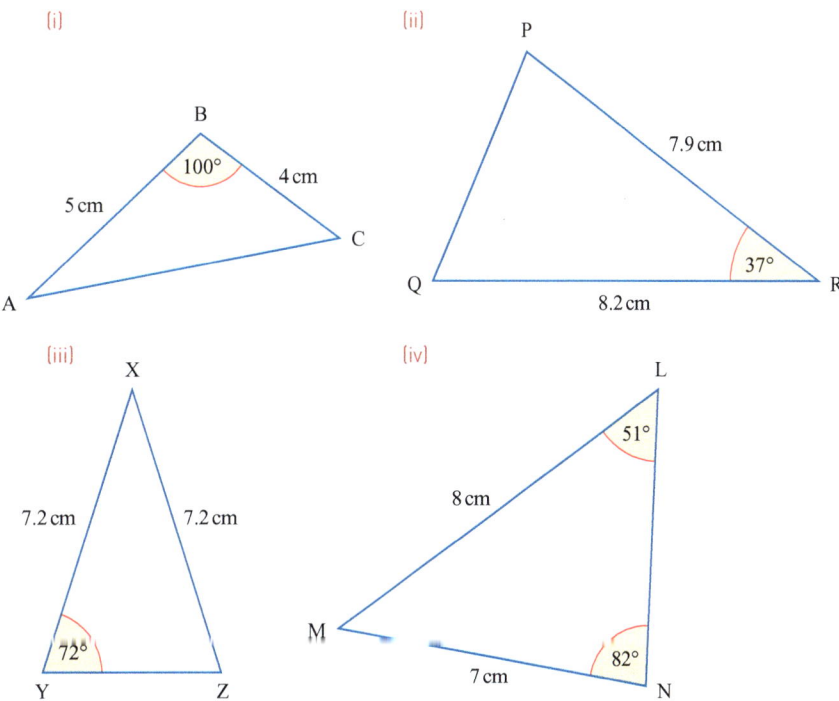

Figure 7.5

② A regular hexagon is made up of six equilateral triangles. Work out the area of a regular hexagon of side 7 cm.

③ A pyramid on a square base has four identical triangular faces which are isosceles triangles with equal sides 9 cm and equal angles 72°.
  (i) Work out the area of a triangular face.
  (ii) Work out the length of a side of the base.
  (iii) Hence work out the total surface area of the pyramid.

④ A tiler wishes to estimate the number of triangular tiles needed to tile an area of 10 m². The dimensions of each tile are shown in the diagram.

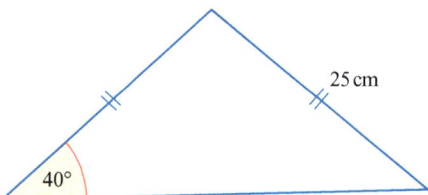

Figure 7.6

  (i) Work out the area of a tile.
The tiler then divides 10 m² by this area and rounds to the next whole number.
  (ii) What result would this give?
  (iii) Explain what is wrong with this estimate.

⑤ A regular tetrahedron has four faces, each of which is an equilateral triangle of side 10 cm. Work out the total surface area of the tetrahedron.

Arcs and sectors

IR ⑥ The area of a rhombus is $\sqrt{48}$ cm². Given also that one of its interior angles is 120°, work out the length of its shortest diagonal.

SA ⑦ A square with sides of length 2 cm has the same area as an equilateral triangle. Work out the side-length of the triangle.

IR ⑧ A circle is drawn inside a square, so that they touch at four points as shown.

A rectangle of dimensions 1 cm × 2 cm is drawn in the corner of the square and touches the circle once. The sides of the rectangle are parallel to the sides of the square. A radius of the circle is drawn to the point where the rectangle meets the circle.
Work out the size of the angle marked θ.

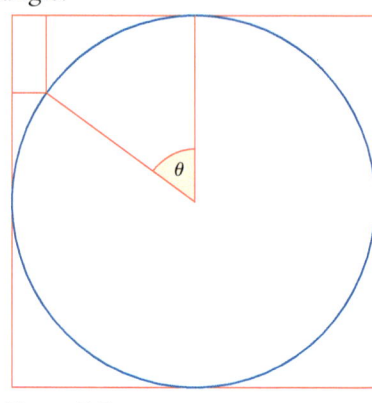

Figure 7.7

2 Arcs and sectors

The length of an *arc* of a circle (Figure 7.8) is calculated by considering it as a fraction of the circumference of the full circle.

$$\text{arc-length} = \frac{\alpha}{360} \times 2\pi r$$

Similarly, the area of a *sector* (Figure 7.9) can be calculated as a fraction of the area of the full circle.

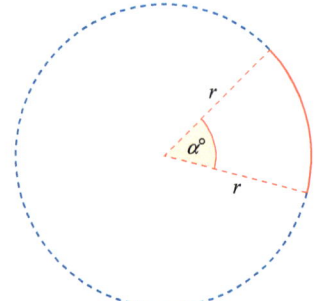

Figure 7.8

$$\text{sector area} = \frac{\alpha}{360} \times \pi r^2$$

A *segment* of a circle is the region between a chord and the edge of the circle (Figure 7.10).

Figure 7.9

The area of a segment can be calculated as area of a triangle subtracted from area of a sector.

> The area of a triangle is $\frac{1}{2}ab\sin C$.

$$\text{segment area} = \frac{\alpha}{360} \times \pi r^2 - \frac{1}{2}r^2 \sin\alpha°$$

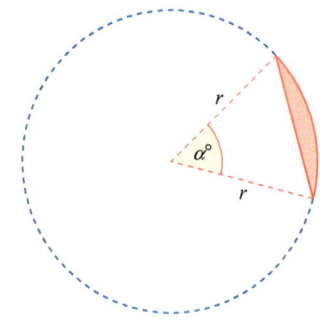

Figure 7.10

Example 7.3

Calculate

(i) the length of arc AB

(ii) the area of sector OAB

(iii) the area of the segment bounded by chord AB and arc AB.

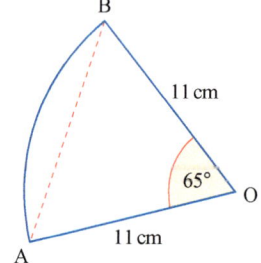

Figure 7.11

Solution

(i)   arc-length $AB = \dfrac{65}{360} \times 2\pi \times 11 = 12.5$ cm (3 s.f.)

(ii) sector area $= \dfrac{65}{360} \times \pi \times 11^2 = 68.6$ cm^2 (3 s.f.)

(iii) triangle area $= \dfrac{1}{2} \times 11^2 \times \sin 65° = 54.8$ cm^2 (3 s.f.)

segment area $= 68.6 - 54.8 = 13.8$ cm^2 (3 s.f.)

Exercise 7B

Give your answers to 3 significant figures unless an exact answer is required.

Note: An **exact** answer is one left as a simplified surd, an improper fraction and may also be in terms of π.

① Calculate the length of each circular arc.

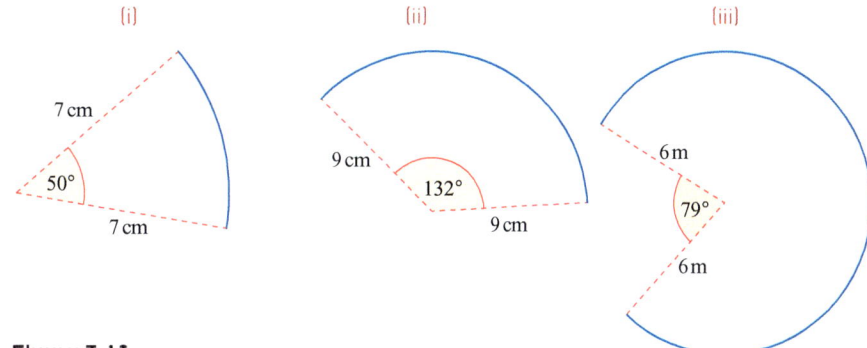

Figure 7.12

② Calculate the area of each sector.

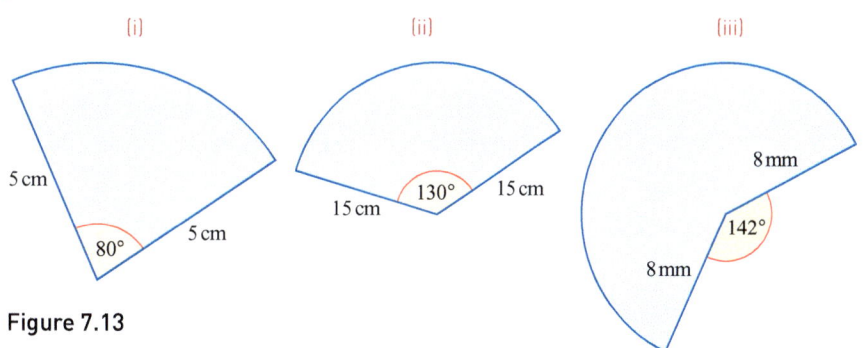

Figure 7.13

③ Calculate the area of each shaded segment.

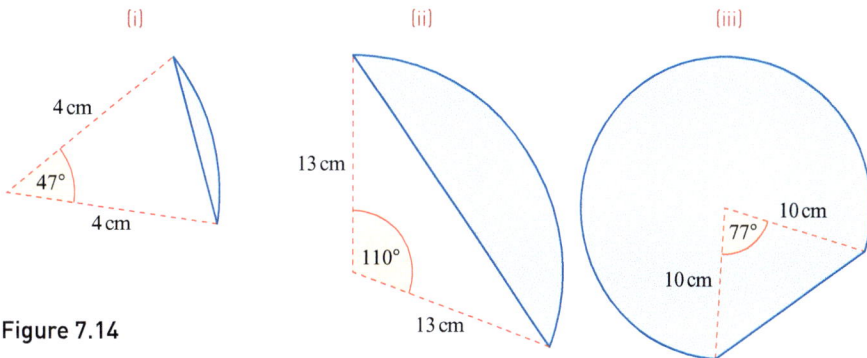

Figure 7.14

The sine rule

SA ④ Find the area and the perimeter of the sector of a circle shown in Figure 7.15.

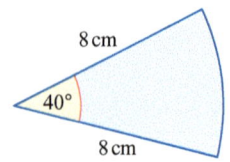

Figure 7.15

IR ⑤ A sector of a circle of radius 11 cm has a perimeter of 30 cm.

Find the angle subtended at the centre of the circle.

IR ⑥ A chord of a circle subtends an angle of 150° at the centre of the circle.

Find, in the form $m : 1$, the ratio

    major segment area : minor segment area

Write m correct to 3 significant figures.

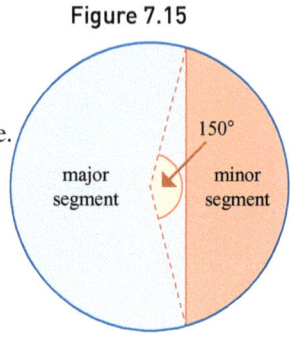

Figure 7.16

IR ⑦ OAD and OBC are sectors of two concentric circles with centre O. OAB and ODC are straight lines.

Shape ABCD is created by removing sector OAD from sector OBC.

OA = 7 cm and AB = 3 cm.

The perimeter of sector OBC is 40 cm.

Calculate

(i) the perimeter of shape ABCD

(ii) the area of shape ABCD.

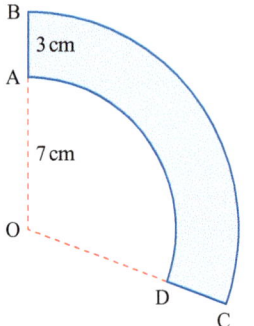

Figure 7.17

The following two trigonometric rules can be used in any triangle, which makes them particularly useful when dealing with scalene triangles.

3 The sine rule

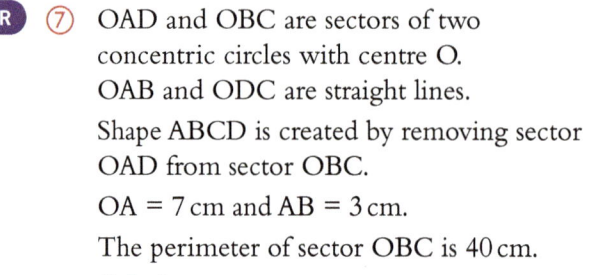

Figure 7.18

You have already seen that for any triangle ABC

$$\text{area} = \tfrac{1}{2}bc\sin A = \tfrac{1}{2}ca\sin B = \tfrac{1}{2}ab\sin C$$

$$\Rightarrow \frac{bc\sin A}{abc} = \frac{ca\sin B}{abc} = \frac{ab\sin C}{abc}$$

$$\Rightarrow \frac{\sin A}{a} = \frac{\sin B}{b} = \frac{\sin C}{c}$$

> **Note**
>
> If a triangle is right-angled then it is much simpler to use the basic trig ratios and/or Pythagoras' theorem. However, the sine and cosine rules are still applicable.

This is one form of the *sine rule* and is the version that is easier to use if you want to work out the size of an angle.

Inverting this gives

$$\frac{a}{\sin A} = \frac{b}{\sin B} = \frac{c}{\sin C}$$

which is better when you need to work out the length of a side.

> **Discussion point**
>
> → Why is the inverted form of the sine rule better when you want to work out the length of a side?

Example 7.4

Work out the length of the side BC in the triangle shown in Figure 7.19.

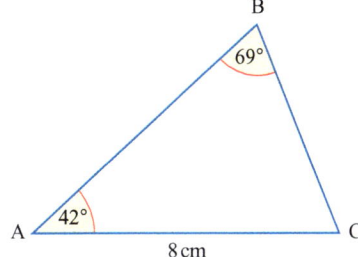

Figure 7.19

> When using the sine rule to work out the size of an angle, you need to be careful because sometimes there are two possible answers, as in Example 7.5. The reason this problem occurs is that for any positive sine ratio there are two possible angles in the range $0°$ to $180°$, except $\sin 90° = 1$. However, this ambiguous case of the sine rule is not assessed in the WJEC Additional Maths examinations.

Solution

Using the sine rule
$$\frac{a}{\sin A} = \frac{b}{\sin B} = \frac{c}{\sin C}$$

$\therefore \quad \dfrac{a}{\sin 42°} = \dfrac{8}{\sin 69°}$

$\Rightarrow \quad a = \dfrac{8 \sin 42°}{\sin 69°}$

$\qquad = 5.733887...$

$\therefore \quad$ side BC $= 5.7$ cm (1 d.p.)

It is advisable to do the calculation entirely on your calculator, and round only the final answer.

Example 7.5

Work out the size of the angle P in the triangle PQR, given that $R = 32°$, $r = 4$ cm and $p = 7$ cm where r and p are the lengths of the sides opposite angles R and P respectively.

Solution

The sine rule for $\triangle$PQR is
$$\frac{\sin P}{p} = \frac{\sin Q}{q} = \frac{\sin R}{r}$$

$\therefore \quad \dfrac{\sin P}{7} = \dfrac{\sin 32°}{4}$

$\Rightarrow \quad \sin P = 0.927\,358\,712$

$\Rightarrow \quad P = 68.0°$ (1 d.p.) or $P = 180° - 68.0° = 112.0°$ (1 d.p.)

Both solutions are possible as indicated in Figure 7.21.

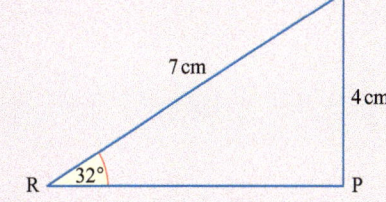

Figure 7.20

> Always check that the second option is a valid solution. Sometimes one of the options gives an impossible triangle. For example, if a solution has an angle sum which is greater than $180°$ then we can reject it. Also, the longest side must be opposite the largest angle, and the shortest side must be opposite the smallest angle.

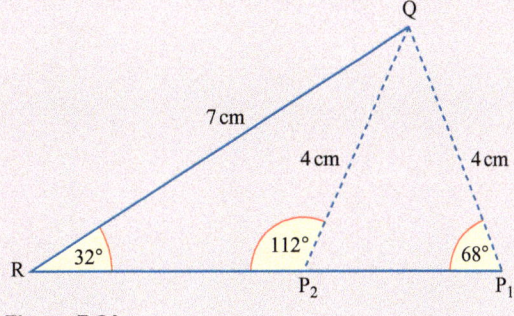

Figure 7.21

The sine rule

> **Note**
>
> You may have met this situation when studying congruent triangles in GCSE maths.
>
> If two triangles have SSS, SAS, ASA (or AAS) or RHS in common then they are congruent. However, if they have ASS in common (as in Example 7.5), then they are not necessarily congruent, as there are two possible triangles.

ACTIVITY 7.1

Figure 7.22 shows triangle XYZ with XY = 6 cm, XZ = 8 cm and ∠XYZ = 78°.

What happens when you use the sine rule to calculate the remaining angles?

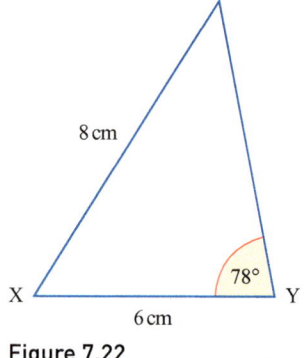

Figure 7.22

Exercise 7C

Where necessary leave answers approximated to 3 significant figures.

① Work out the length x in each of these triangles.

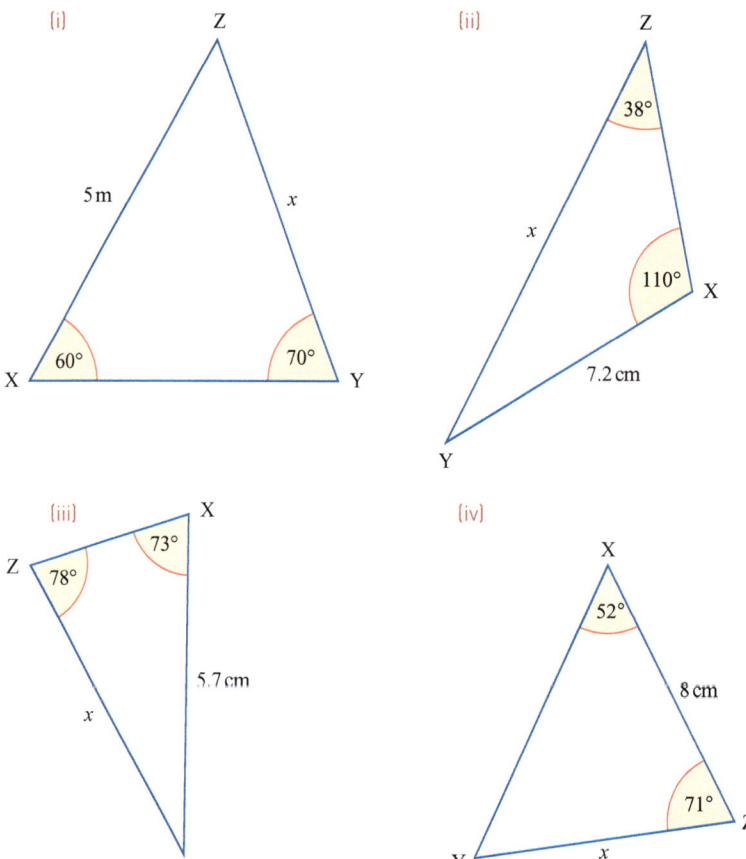

Figure 7.23

② Work out the size of the angle θ in each of these triangles.

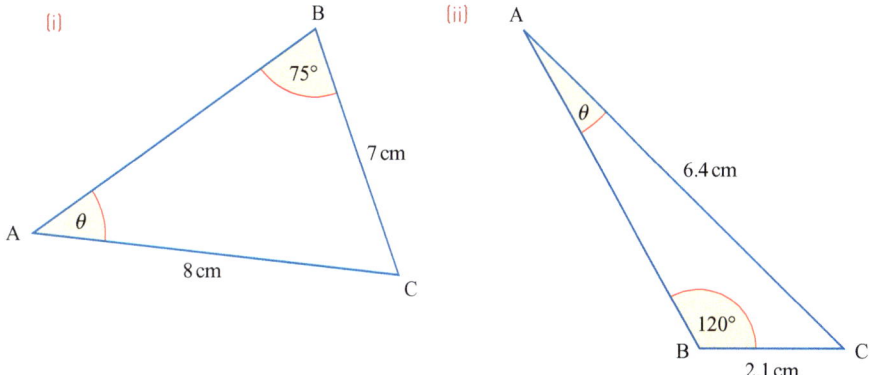

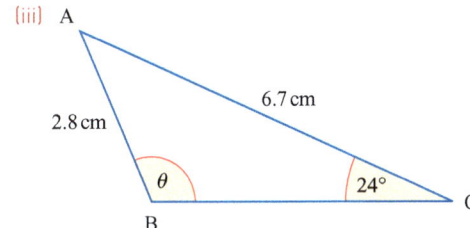

Figure 7.24

SA ③ Work out the size of the angle marked x in the quadrilateral shown.

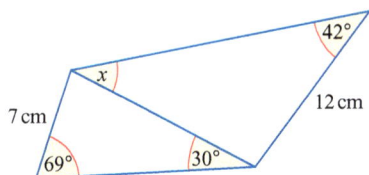

Figure 7.25

SA ④ Carys walks on a bearing of 132° for 4 km.

She then changes direction and walks on a bearing of 017° until she is due east of her starting position.

How far is she from her starting position?

SA ⑤ The angles of a kite are 122°, 102°, 102° and 34°.

The diagonal which lies along the kite's line of symmetry is 12 cm in length.

Work out the lengths of each of the kite's four sides.

IR ⑥ Ffion and Gwen are at point P.

Point Q is due north of point P, as shown in Figure 7.26.

They disagree about the shortest route from P to Q.

Ffion walks on a bearing of 330°, and then changes to a bearing of 040°, which takes her straight to Q.

Gwen walks 3 km on a bearing of 020°, after which she walks on a bearing of 300°, which then takes her straight to Q.

Who took the shorter route?

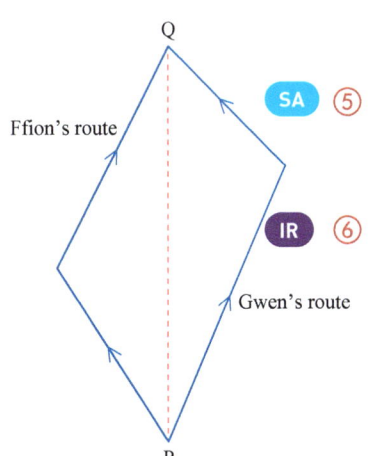

Figure 7.26

4 The cosine rule

Sometimes it is not helpful to use the sine rule with the information you have about a triangle, for example, if you know all three side lengths but none of the angles.

Like the sine rule, the cosine rule can be applied to any triangle, and again there are equivalent versions.

Use this version to work out a side length.

$$a^2 = b^2 + c^2 - 2bc\cos A$$

Use this version to work out the size of an angle.

$$\cos A = \frac{b^2 + c^2 - a^2}{2bc}$$

Proof

For the triangle ABC, line CD is perpendicular to side AB as shown in Figure 7.27.

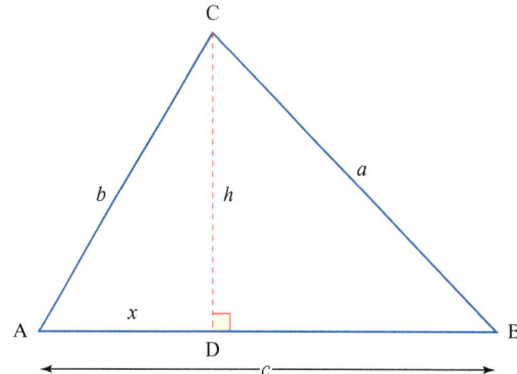

Figure 7.27

In $\triangle ACD$ (This is shorthand notation for 'triangle'.)

$$b^2 = x^2 + h^2 \qquad \qquad ①$$

(Pythagoras' theorem.)

and $\cos A = \dfrac{x}{b}$ so $x = b\cos A$ ②

In $\triangle BCD$ (Pythagoras' theorem.)

$$a^2 = (c - x)^2 + h^2$$

$\Rightarrow \quad a^2 = c^2 - 2cx + x^2 + h^2$

$\Rightarrow \quad a^2 = c^2 - 2cx + b^2 \qquad$ using ①

$\Rightarrow \quad a^2 = c^2 - 2cb\cos A + b^2 \qquad$ using ②

$\Rightarrow \quad a^2 = b^2 + c^2 - 2bc\cos A \qquad$ (as required).

Rearranging this gives

$$2bc\cos A = b^2 + c^2 - a^2$$

$\Rightarrow \quad \cos A = \dfrac{b^2 + c^2 - a^2}{2bc}$

which is the second form of the cosine rule.

> **Note**
>
> Starting with a perpendicular from a different vertex would give the following similar results.
>
> $b^2 = a^2 + c^2 - 2ac\cos B$ and $\cos B = \dfrac{a^2 + c^2 - b^2}{2ac}$
>
> $c^2 = a^2 + b^2 - 2ab\cos C$ and $\cos C = \dfrac{a^2 + b^2 - c^2}{2ab}$

Example 7.6

Work out the length of the side AB in the triangle shown in Figure 7.28.

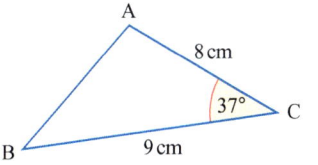

Figure 7.28

Solution

$c^2 = a^2 + b^2 - 2ab\cos C$

$c^2 = 9^2 + 8^2 - 2 \times 9 \times 8 \times \cos 37°$

$ = 29.996$

$AB = 5.48\,\text{cm}$ (3 s.f.)

> There are two common errors when using this formula.
>
> - In a non-calculator paper, evaluate the three terms a^2, b^2 and $2ab\cos C$ separately. A common error is to calculate $a^2 + b^2 - 2ab$ and then multiply by $\cos C$. However, these questions are usually in calculator papers, in which case the whole calculation can be typed into a scientific calculator – this will deal with the priority of operations correctly.
>
> - Another common error is to forget to square root after calculating $a^2 + b^2 - 2ab\cos C$.

Example 7.7

Work out the size of the angle P in the triangle shown in Figure 7.29.

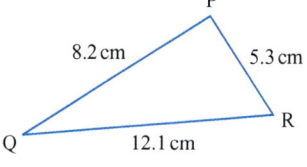

Figure 7.29

Solution

The cosine rule for this triangle can be written as

$\cos P = \dfrac{q^2 + r^2 - p^2}{2qr}$

$\cos P = \dfrac{5.3^2 + 8.2^2 - 12.1^2}{2 \times 5.3 \times 8.2}$

$\cos P = -0.588$

$P = 126.0°$ (1 d.p.)

The cosine rule

Exercise 7D

Where necessary leave answers approximated to 3 significant figures.

① Work out the length x in each of these triangles.

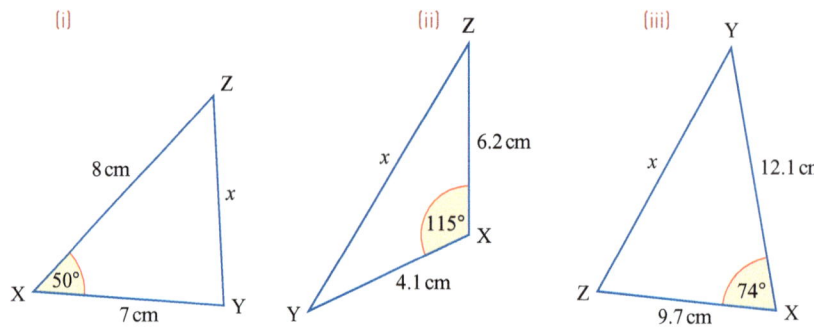

Figure 7.30

② Work out the size of the angle θ in each of the following triangles.

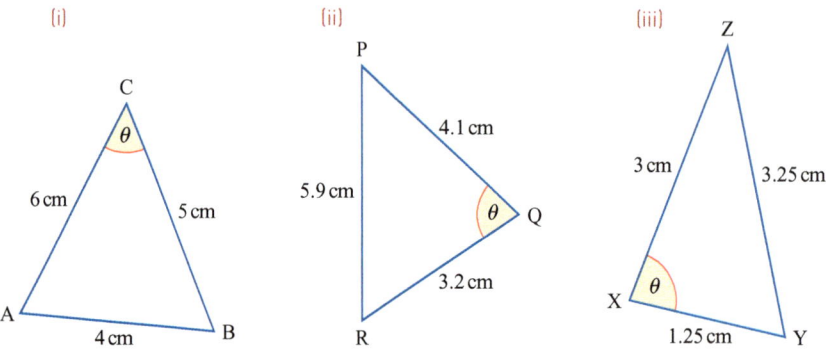

Figure 7.31

SA ③ The diagonals of a parallelogram have lengths of 12 cm and 18 cm and the angle between them is 72°. Work out the lengths of the sides of the parallelogram.

SA ④ Figure 7.32 shows a quadrilateral ABCD with AB = 8 cm, BC = 6 cm, CD = 7 cm, DA = 5 cm and ∠ABC = 90°.
Calculate
  (i) AC
  (ii) ∠ADC.

Figure 7.32

IR ⑤ Figure 7.33 shows two circles. One has centre A and a radius of 8 cm. The other has centre B and a radius of 10 cm. AB = 12 cm and the circles intersect at P and Q.
Calculate ∠PAB.

SA ⑥ A parallelogram has sides of length 5 cm and 10 cm, and an angle of 130°. Work out the length of the longest diagonal.

IR ⑦ A triangle has sides of length 6 cm, 7 cm and 11 cm. Work out its area.

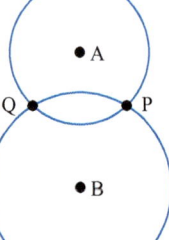

Figure 7.33

SA ⑧ Aled walks 7 km on a bearing of 054°.

He then walks 5 km on a bearing of 122°.

How far is he from his starting point?

IR ⑨ In triangle ABC, AB = 8 cm, BC = 5 cm and ∠BAC = 35°.

Use the cosine rule to work out the possible lengths of AC.

5 Using the sine and cosine rules together

> **Note**
> Three angles are not independent measurements, as a third angle can be calculated from the other two.

When solving any triangle, three independent measurements are required.

Given *three side-lengths*, use the cosine rule to work out the size of an angle.

Given *two side-lengths and an included angle*, use the cosine rule to work out the length of the third side.

Given *two side-lengths and an angle (not included)*, use the sine rule to work out the size of another angle from which you can calculate the included angle, and you can then use the cosine rule to work out the missing length. This situation can sometimes produce two possible solutions.

Given *two angles and one side-length*, use the sine rule to work out another side-length. If the given side-length is between the two angles, then first calculate the size of the third angle using the angle sum of a triangle.

Once a fourth independent measurement of a triangle has been calculated, then the other two can be calculated using either the cosine rule or the sine rule.

Example 7.8

Figure 7.34 shows the positions of three towns, Aldbury, Bentham and Chorton.

Bentham is 8 km from Aldbury on a bearing of 037° and Chorton is 9 km from Bentham on a bearing of 150°.

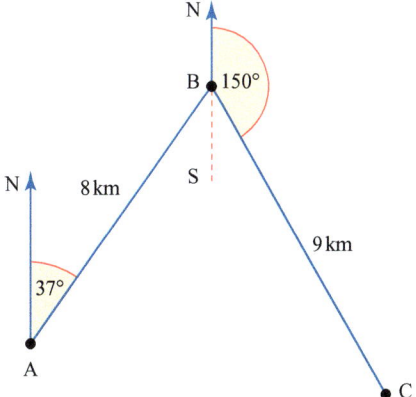

Figure 7.34

Work out

(i) the size of the angle ABC

(ii) the distance of Chorton from Aldbury (to the nearest 0.1 km)

(iii) the bearing of Chorton from Aldbury (to the nearest 1°).

Using the sine and cosine rules together

Solution

(i) ∠ABS = 37° (alternate angles)
and ∠SBC = 30° (adjacent angles on a straight line)
so ∠ABC = 67°.

(ii) Using the cosine rule
$$b^2 = a^2 + c^2 - 2ac \cos B$$
$$= 9^2 + 8^2 - 2 \times 9 \times 8 \cos 67°$$
$$= 88.7347...$$
$$b = 9.4199...$$

> Don't clear this from your calculator as you will need it later.

Chorton is 9.4 km (1 d.p.) from Aldbury.

(iii) Using the sine rule
$$\frac{\sin A}{a} = \frac{\sin B}{b}$$
$$\frac{\sin A}{9} = \frac{\sin 67°}{9.4199...}$$
$$\sin A = 0.879\,47...$$
$$A = 61.57...°$$

The bearing of Chorton from Aldbury is 099°.

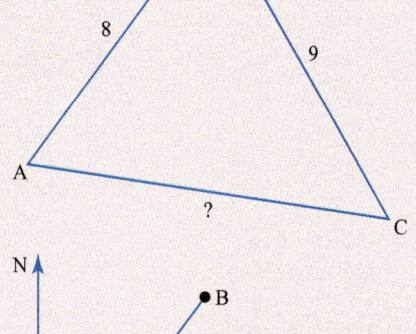
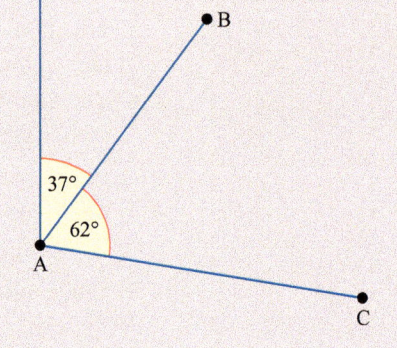

Figure 7.35

> **Discussion point**
> → The other value of A that gives sin A = 0.879 47... is 118.42...°
> Why does this not give an alternative solution to this problem?

Example 7.9

A triangular plot of land has sides of length 70 m, 80 m and 95 m.

Work out its area in hectares. (1 hectare is 10 000 m².)

Solution

First draw a sketch and label the sides.

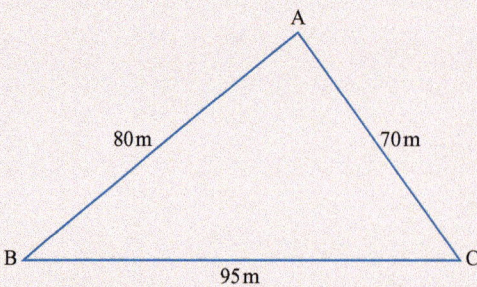

Figure 7.36

You can now see that the first step is to work out the size of one of the angles, and this will need the cosine rule.

$$\cos A = \frac{b^2 + c^2 - a^2}{2bc}$$

$$\cos A = \frac{70^2 + 80^2 - 95^2}{2 \times 70 \times 80}$$

$$= \frac{13}{64}$$

$$\Rightarrow A = 78.28°$$

$$\text{Area} = \frac{1}{2} bc \sin A$$

$$\text{Area} = \frac{1}{2} \times 70 \times 80 \times \sin 78.28°$$

$$= 2742 \, \text{m}^2$$

$$= 0.27 \, \text{hectares (2 d.p.)}$$

Exercise 7E

Where necessary leave answers approximated to 3 significant figures.

1. The hands of a clock have lengths 6 cm and 8 cm.

   Work out the distance between the tips of the hands at 8 p.m.

2. From a lighthouse L, a ship A is 4 km away on a bearing of 340° and a ship B is 5 km away on a bearing of 065°.

   Work out the distance AB.

3. When I am at a point X, the angle of elevation of the top (T) of a vertical tree is 27°, but if I walk 20 m towards the tree along horizontal ground, to point Y, the angle of elevation is then 47°.

   (i) Work out the distance TY.

   (ii) Work out the height of the tree.

4. Two adjacent sides of a parallelogram have lengths 9.3 cm and 7.2 cm, and the shorter diagonal is of length 8.1 cm.

   (i) Work out the sizes of the angles of the parallelogram.

   (ii) Work out the length of the other diagonal of the parallelogram.

5. A yacht sets off from A and sails for 5 km on a bearing of 067° to a point B so that it can clear the headland before it turns onto a bearing of 146°. It then stays on that course for 8 km until it reaches a point C.

   (i) Work out the distance AC.

   (ii) Work out the bearing of C from A.

6. Two ships leave the docks, D, at the same time. *Princess Pearl*, P, sails on a bearing of 160° at a speed of 18 km h^{-1}, and *Regal Rose*, R, sails on a bearing of 105°. After 2 hours the angle DRP is 80°.

   (i) Work out the distance between the ships at this time.

   (ii) Work out the speed of the *Regal Rose*.

Mensuration

SA ⑦ The diagram in Figure 7.37 represents a simplified drawing of the timber cross-section of a roof.

(i) Work out the lengths of the struts BD and EG.

(ii) Work out the length DE.

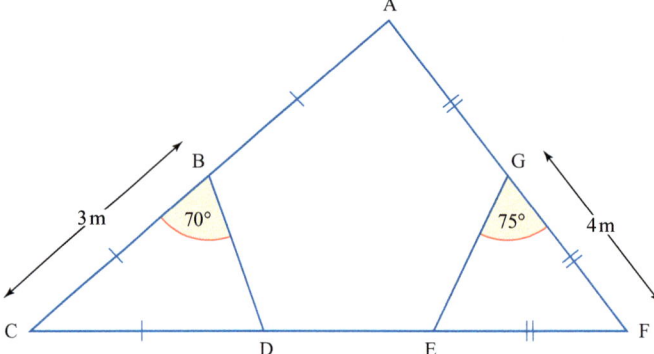

Figure 7.37

SA ⑧ Dylan and Aziz cycle home from school.

Dylan cycles due east for 4 km, and Aziz cycles due south for 3 km and then for 2 km on a bearing of 125°.

How far apart are their homes?

6 Mensuration

- Area of triangle $= \frac{1}{2} \times$ base $\times$ height
- Area of parallelogram = base $\times$ height
- Area of trapezium $= \frac{1}{2} \times$ sum of parallel sides $\times$ perpendicular distance between them
- Circumference of circle $= \pi d = 2\pi r$
- Area of circle $= \pi r^2$
- Volume of prism = area of cross-section $\times$ length
- Volume of sphere $= \frac{4}{3}\pi r^3$
- Surface area of sphere $= 4\pi r^2$
- Volume of pyramid $= \frac{1}{3} \times$ base area $\times$ height
- Volume of cone $= \frac{1}{3}\pi r^2 h$
- Curved surface area of cone $= \pi r l$

Example 7.10

A sphere has diameter 10 cm.

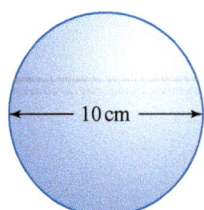

Figure 7.38

(i) Find its volume. (ii) Find its surface area.

Solution

The radius is 5 cm.

(i) $\text{volume} = \frac{4}{3}\pi r^3$

$= \frac{4}{3} \times \pi \times 5^3$

$= 524 \text{ cm}^3$ (3 s.f.)

(ii) $\text{surface area} = 4\pi r^2$

$= 4 \times \pi \times 5^2$

$= 314 \text{ cm}^2$ (3 s.f.)

Example 7.11

A cone has height 15 cm and a base of diameter 16 cm.

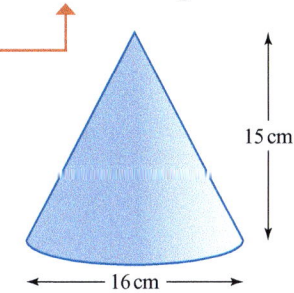

Unless a question specifies otherwise, assume the vertex (or apex) of a cone is directly above the centre of the base, and that the base is circular. Such a cone is sometimes referred to as a right circular cone.

Figure 7.39

(i) Find its exact volume.

(ii) Find its exact surface area.

An exact answer is not an approximation. If a calculation involves the use of π then avoid substituting a decimal approximation, for example 3.14. Instead, treat π in the same way you would x in an algebraic expression.

Other examples of exact answers include fractions and surds. For example, answers of $\frac{17}{6}$ and $\sqrt{2}$ are exact, but 2.833 (3 d.p.) and 1.414 (4 s.f.) are not.

Solution

The radius is 8 cm.

(i) $\text{volume} = \frac{1}{3}\pi r^2 h$

$= \frac{1}{3} \times \pi \times 8^2 \times 15$

$= 320\pi \text{ cm}^3$

(ii) $\text{slanted edge} = \sqrt{8^2 + 15^2}$

$= 17 \text{ cm}$

The surface area of a cone comprises the curved surface area and the flat circular base.

$\text{surface area} = \pi r l + \pi r^2$

$= \pi \times 8 \times 17 + \pi \times 8^2$

$= 200\pi \text{ cm}^2$

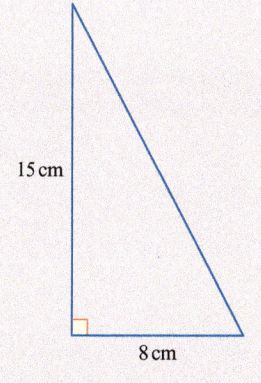

Figure 7.40

Mensuration

Example 7.12

A pyramid has a square base of sides 10 cm.

The vertex of the pyramid is 12 cm above the centre of the base.

(i) Find its volume.

(ii) Find its surface area.

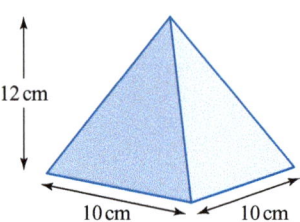

Figure 7.41

Solution

(i) volume = $\frac{1}{3}$ × base area × height

$= \frac{1}{3} \times 10^2 \times 12$

$= 400 \, \text{cm}^3$

(ii)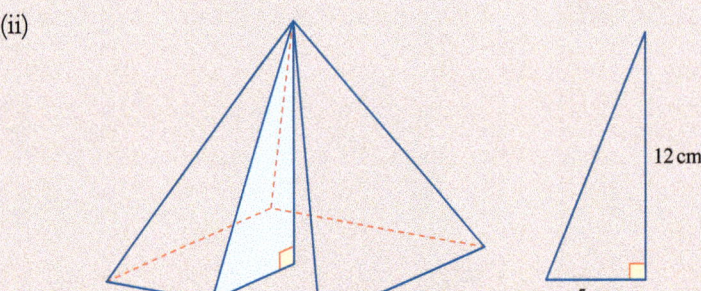

Figure 7.42

The hypotenuse of the shaded triangle is the bisector of the slanted face.

It has length $\sqrt{5^2 + 12^2} = 13$ cm.

So the area of each slanted face is $\frac{1}{2} \times 10 \times 13 = 65 \, \text{cm}^2$.

So the total surface area is $10^2 + 4 \times 65 = 360 \, \text{cm}^2$.

Example 7.13

A cone with a base radius of 6 cm and a height of 3π cm has the same volume as a pyramid with a square base of side 2π cm. What is the height of the pyramid?

Solution

Let h = height of pyramid.

Volume of pyramid = volume of cone

$\therefore \quad \frac{1}{3}(2\pi)^2 h = \frac{1}{3}\pi 6^2 \times 3\pi$

$\Rightarrow \quad \frac{4\pi^2 h}{3} = \frac{108\pi^2}{3}$

$\Rightarrow \quad 4h = 108$

$\Rightarrow \quad h = 27$ cm

Exercise 7F

① Calculate the volume of each of these spheres, writing your answers correct to 3 significant figures.
   (i) Radius = 12 cm
   (ii) Radius = 10 m
   (iii) Diameter = 19 mm
   (iv) Diameter = 23 cm

② Calculate the surface area of each of these spheres, writing your answers correct to 3 significant figures.
   (i) Radius = 7 cm
   (ii) Radius = 12 m
   (iii) Diameter = 16 mm
   (iv) Diameter = 37 cm

③ A sphere has a diameter of 12 cm. Calculate its exact volume and surface area.

④ Calculate the volume of each of these cones, writing your answers correct to 3 significant figures.
   (i) Height 7 cm and base radius 15 cm
   (ii) Height 9 m and base radius 11 m
   (iii) Height 10 mm and base diameter 10 mm
   (iv) Height 12 cm and base diameter 37 cm

⑤ Calculate the surface area of each of these cones, writing your answers correct to 3 significant figures.
   (i) Height 8 cm and base radius 18 cm
   (ii) Height 10 m and base radius 12 m
   (iii) Height 9 mm and base diameter 9 mm
   (iv) Height 16 cm and base diameter 25 cm

⑥ A cone has height 24 cm and a base radius of 10 cm. Calculate its exact volume and surface area.

⑦ A pyramid has a square base of sides 12 cm.
   The vertex of the pyramid is 15 cm above the centre of the base.
   (i) Calculate its volume.
   (ii) Calculate its surface area.

⑧ The toy shown in Figure 7.43 comprises a cone and a hemisphere joined at their bases.
   The base radius of each shape is 8 cm.
   The height of the whole shape is 24 cm.
   (i) Calculate its volume.
   (ii) Calculate its surface area.

   A hemisphere is half of a sphere.

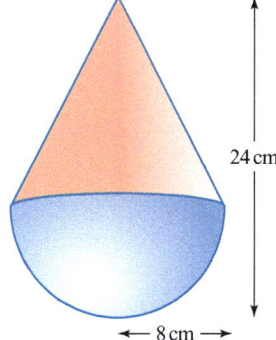

Figure 7.43

⑨ A sphere has volume 100 cm³. Calculate its surface area.

7 Problems in three dimensions

Figure 7.44

Discussion point

→ An aircraft flying between two places at the same latitude doesn't usually follow a route along the line of latitude. Why?

When you are solving three-dimensional problems it is important to draw good diagrams (although you will not be assessed on this in the exam, a clear diagram does benefit understanding). There are two types:

- representations of three-dimensional objects
- true shape diagrams of two-dimensional sections within a three-dimensional object.

Representations of three-dimensional objects

Figures 7.45 and 7.46 illustrate ways in which you can draw a clear diagram.

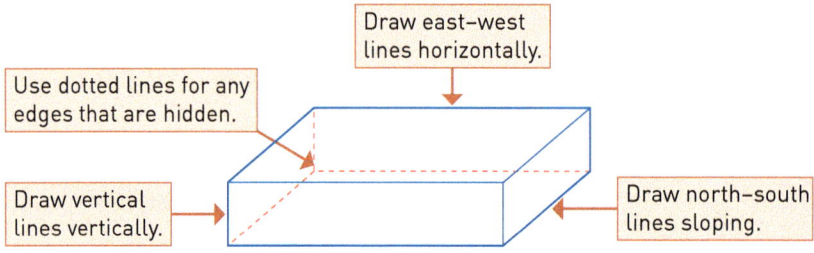

Figure 7.45

> Open up the diagram as much as possible by choosing a suitable direction for your *north–south* axis; (a) is clearer than (b).

(a)

(b)

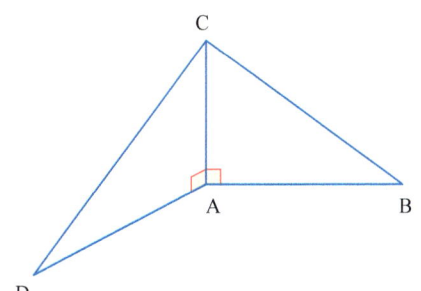
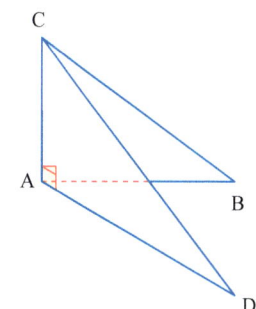

Figure 7.46

True shape diagrams

In a two-dimensional representation of a three-dimensional object, right angles do not always appear to be 90°, so draw as many true shape diagrams as necessary.

For example, if you need to do calculations on the triangular cross-section BCD in Figure 7.47(a), you should draw the triangle so that the right angle really does look 90° as in Figure 7.47(b).

(a)

(b)

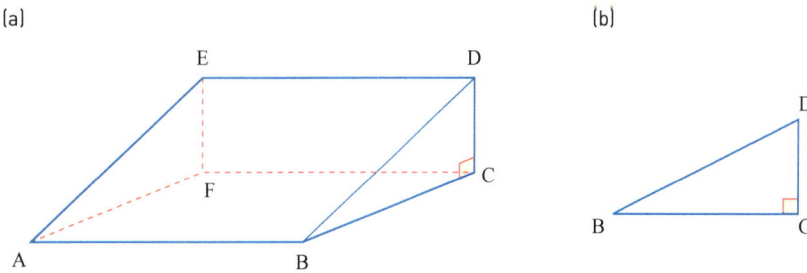

Figure 7.47

8 Lines and planes in three dimensions

A *plane* is a flat surface (not necessarily horizontal).

A *line of greatest slope* of a sloping plane is a line of greatest gradient, i.e. the line that a ball would follow if allowed to roll down it. This is shown in Figure 7.48.

> **Discussion point**
> → Give an example of a sloping plane from everyday life.

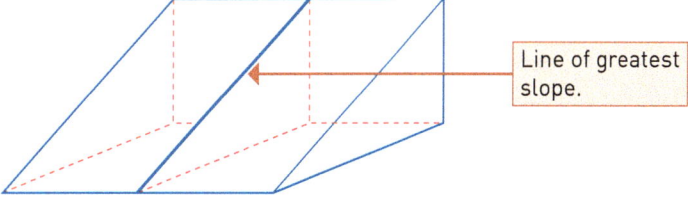

Line of greatest slope.

Figure 7.48

Lines and planes in three dimensions

In three-dimensional problems you need to be aware of the relationships between lines and planes.

Two lines

In two dimensions, two lines either meet (when extended if necessary), or they are parallel.

In three dimensions, there is a third option: they are *skew*, as in Figure 7.49.

The road under the bridge and the road over the bridge are skew lines.

Figure 7.49

A line and a plane

In three dimensions there are three options, as shown in Figure 7.50.

(a) The line and the plane are *parallel*. A curtain rail is *parallel* to the floor.
(b) The line meets the plane at a *single point*. When you are writing, your pen meets the paper at a *single point*.
(c) The line *lies in* the plane. When you put your pen down, your pen *lies in* the plane of the paper.

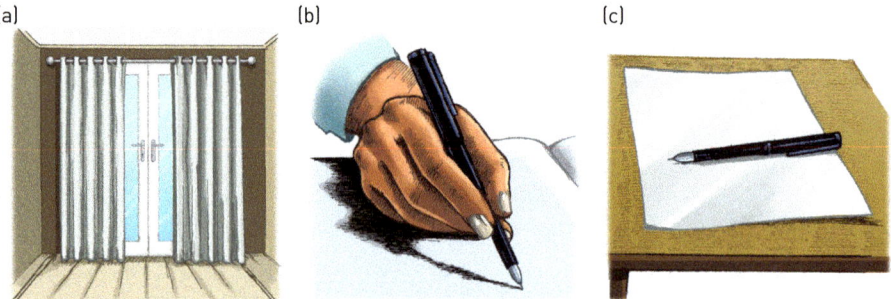

Figure 7.50

Angle between a line and a plane

Draw a perpendicular from the line to the plane.

Line PQ meets the plane ABCD at Q.

PR is perpendicular to the plane.

QR is in the plane.

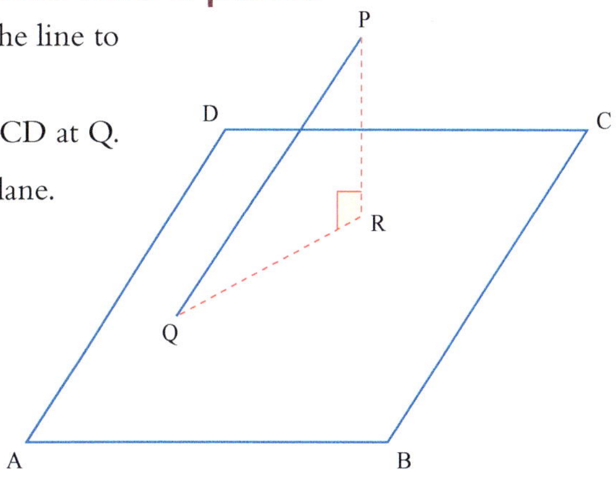

Figure 7.51

The angle between the line and the plane is angle PQR.

Two planes

In three dimensions there are two options.

(a) The two planes are *parallel*. Opposite walls of a room are usually parallel.
(b) The two planes meet *in a line*. The ceiling meets each wall of a room *in a line*. An open gate and a wall meet *in a line*.

Discussion point
→ Give other examples of these cases.

Figure 7.52

Angle between two planes

Identify the line where the planes meet.

Draw a line in each plane that is perpendicular to the line where the planes meet.

The angle between these two lines is the angle between the planes.

Planes ABCD and APQD meet along AD.

The dashed lines are each perpendicular to AD.

x is the angle between the planes.

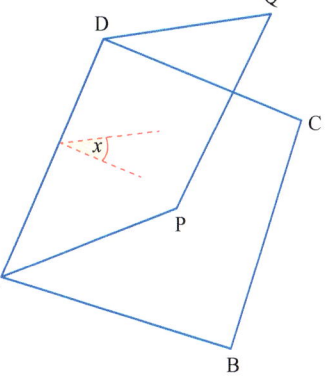

Figure 7.53

Lines and planes in three dimensions

Example 7.14

Figure 7.54 shows a wedge ABCDEF with AB = 8 cm, BC = 6 cm and CD = 2 cm. The angle BCD is 90°.

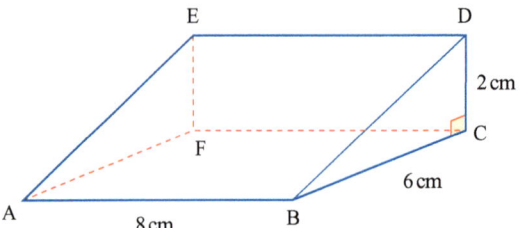

Figure 7.54

Work out

(i) the length AC

(ii) the length AD

(iii) the size of the angle between DA and ABCF

(iv) the size of the angle between ABDE and ABCF.

Solution

(i) From Figure 7.55(a)

$AC^2 = 8^2 + 6^2$ (Pythagoras)

$\Rightarrow AC = 10\,cm$

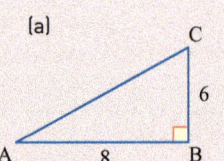

(ii) From Figure 7.55(b)

$AD^2 = AC^2 + 2^2$ (Pythagoras)

$\Rightarrow AD = 10.2\,cm$ (1 d.p.)

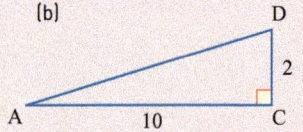

> DA and ABCF meet at A. DC is perpendicular to ABCF.

(iii) From Figure 7.55(b), the angle between DA and ABCF is ∠DAC.

$\tan \angle DAC = \dfrac{2}{10}$

$\Rightarrow \angle DAC = 11.3°$ (1 d.p.)

(iv) From Figure 7.55(c), the angle between ABDE and ABCF is ∠DBC.

$\tan \angle DBC = \dfrac{2}{6}$

$\Rightarrow \angle DBC = 18.4°$ (1 d.p.)

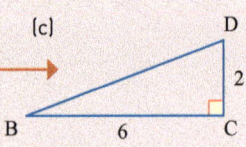

Figure 7.55

> ABDE and ABCF meet along AB. BD is perpendicular to AB. BC is perpendicular to AB.

Example 7.15

RWC

Figure 7.56 shows a straight level road AB, 400 m long. A vertical radio mast XY stands some distance from the road, and the bottom of the mast, X, is on the same level as the road. The angle of elevation of Y from A is 30°, ∠XAB = 25° and ∠AXB = 90°.

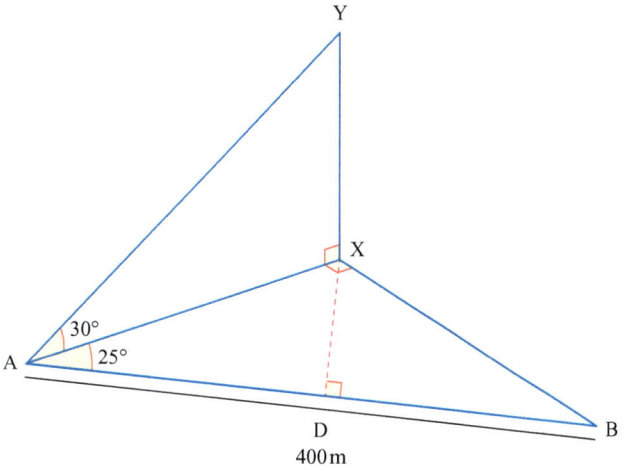

Figure 7.56

Calculate

(i) the distance AX

(ii) the height of the mast

(iii) the distance of X from the road.

Give your answers to 3 significant figures.

Solution

(i) From Figure 7.57(a)

$$\frac{AX}{400} = \cos 25°$$

⇒ AX = 362.523…

⇒ The distance AX = 363 m.

(a)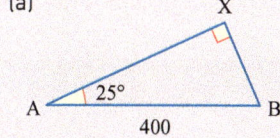

(ii) From Figure 7.57(b)

$$\frac{XY}{362.523…} = \tan 30°$$

⇒ XY = 209.302…

⇒ The height of the mast XY = 209 m.

(b)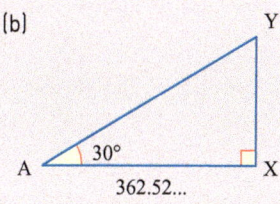

(iii) From Figure 7.57(c)

$$\frac{DX}{362.523…} = \sin 25°$$

⇒ DX = 153.208…

⇒ The distance of X from the road = 153 m.

(c)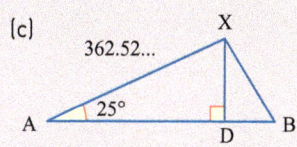

Figure 7.57

Lines and planes in three dimensions

Example 7.16

The pyramid VABCD has square horizontal base ABCD.

The vertex, V, is directly above the centre, X, of the base.

M is the midpoint of BC.

AB = 8 metres and VX = 15 metres.

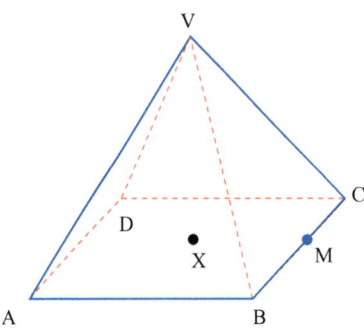

Figure 7.58

Work out the angle between the planes ABCD and VBC.

Solution

The planes meet along BC.

MX and VM are both perpendicular to BC.

Angle VXM is 90°.

XM = 8 ÷ 2
  = 4 m

$\tan VMX = \dfrac{15}{4}$

angle VMX = 75.1°   (1 d.p.)

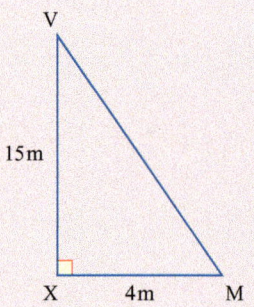

Figure 7.59

Example 7.17

The cuboid has a square base ABCD of side 8 cm and a height of 4 cm.

M is the midpoint of AC.

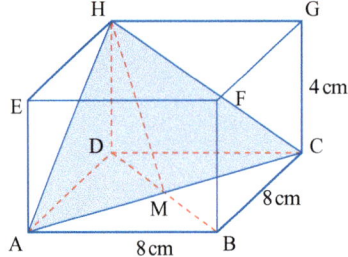

Figure 7.60

(i) Calculate the exact length of DM.

(ii) Work out the angle between the planes ABCD and ACH.

Solution

(i) DM = half the length of the diagonal of the square base

$$= \frac{1}{2}\sqrt{8^2 + 8^2}$$

$$= 4\sqrt{2} \text{ cm}$$

(ii) The angle required is ∠HMD.

$$\tan \text{HMD} = \frac{\text{HD}}{\text{DM}}$$

$$= \frac{4}{4\sqrt{2}}$$

The required angle is 35.3°.

Pythagoras' theorem in three dimensions

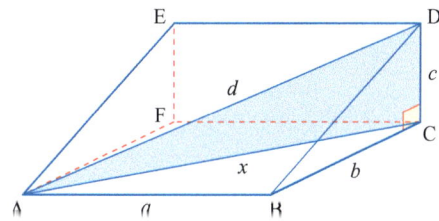

Figure 7.61

> **ACTIVITY 7.2**
>
> Look for integer values of a, b, c and d such that $a^2 + b^2 + c^2 = d^2$
> You could start with $3^2 + 4^2 = 5^2$ and then use $5^2 + 12^2 = 13^2$
> Can you find at least two examples of values of a, b, c and d?

In Figure 7.61, the base is rectangular, so using Pythagoras' theorem in two dimensions

$$a^2 + b^2 = x^2$$

The triangle ACD has a right angle at C, giving

$$x^2 + c^2 = d^2$$

Substituting for x^2 from the first equation gives

$$a^2 + b^2 + c^2 = d^2$$

This is the 3-D version of Pythagoras' theorem.

Example 7.18

ABCDEFGH is a cuboid with side-lengths as shown in the diagram.

Calculate the length of the diagonal AF.

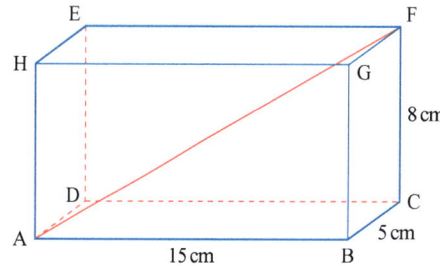

Figure 7.62

Solution

$$\text{AF} = \sqrt{15^2 + 5^2 + 8^2}$$

$$= \sqrt{314}$$

$$= 17.7 \text{ cm } (3 \text{ s.f.})$$

Lines and planes in three dimensions

Use of the sine and cosine rules in 3-D problems

Example 7.19

ABCDEFGH is a cuboid with side-lengths as shown in the diagram.

(i) Calculate the size of angle HDF.

(ii) Hence, calculate the size of angle DHF.

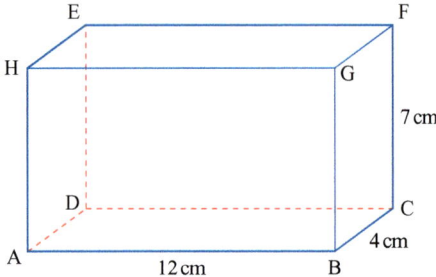

Figure 7.63

Solution

Consider triangle HDF.

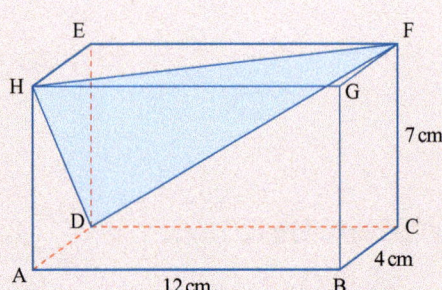

(i) $HD = \sqrt{4^2 + 7^2} = \sqrt{65}$

$FD = \sqrt{12^2 + 7^2} = \sqrt{193}$

$HF = \sqrt{4^2 + 12^2} = \sqrt{160}$

Figure 7.64

Using the cosine rule: $\cos HDF = \dfrac{HD^2 + FD^2 - HF^2}{2 \times HD \times FD}$

$\cos HDF = \dfrac{65 + 193 - 160}{2 \times \sqrt{65} \times \sqrt{193}}$

$\cos HDF = 0.437$

$HDF = 64.1°$ (1 d.p.)

(ii) Using the sine rule: $\dfrac{\sin H}{FD} = \dfrac{\sin D}{HF}$

$\dfrac{\sin H}{\sqrt{193}} = \dfrac{\sin 64.1}{\sqrt{160}}$

$\sin H = 0.988$

$DHF = 81.0°$ (1 d.p.)

Exercise 7G

1. The cube ABCDEFGH shown in the diagram has sides of length 10 cm.

   Calculate
   (i) the length AC
   (ii) the length AG
   (iii) the angle GAC.

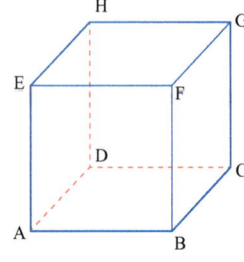

Figure 7.65

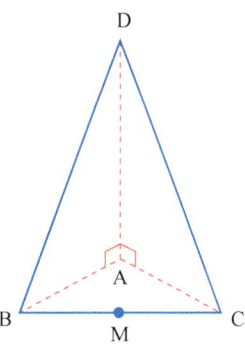

Figure 7.66

2. Figure 7.66 represents a pyramid ABCD with a horizontal base ABC.

AB = AC = 5 cm and BD = CD = 13 cm.

BC = 8 cm.

D is vertically above A and ∠BAD = ∠CAD = 90°.

M is the midpoint of BC.

Calculate

(i) the length AM
(ii) the angle BCD
(iii) the angle between the planes BCA and BCD.

3. Figure 7.67 shows a wedge ABCDEF which has been made to hold a door open.

AB = 5 cm, BC = 12 cm and FC = 4 cm.

Calculate

(i) the angle FBC
(ii) the length AC
(iii) the angle between the line FA and the plane ABCD.

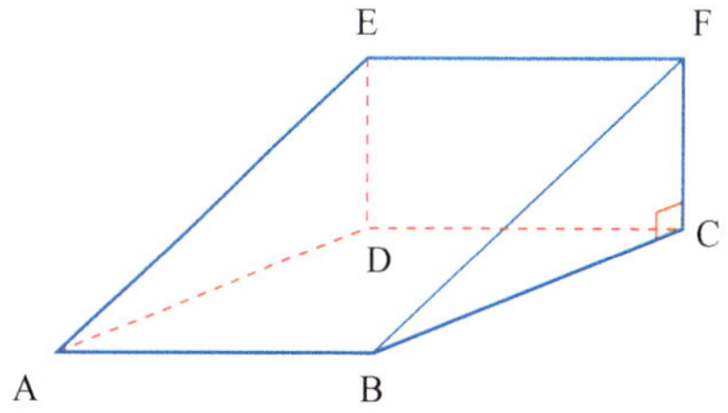

There is a gap of 2 cm between the door and the floor.

(iv) How far along BF will the base of the door meet the wedge?

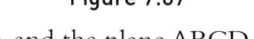
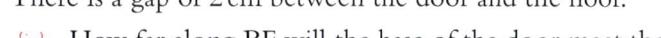

Figure 7.67

4. A, B and C are points on a horizontal plane.

A is 75 m from C on a bearing of 210° and the bearing of B from C is 120°. The bearing of B from A is 075°.

From A, the angle of elevation of the top T of a vertical tower at C is 42°.

Calculate

(i) the distance BC
(ii) the height of the tower
(iii) the angle of elevation of T from B.

5. C is the foot of a vertical tower CT 28 m high.

A and B are points in the same horizontal plane as C, and CA = CB.

P is the point on AB that is nearest to C.

The angle of elevation of the top of the tower from P is 40° and ∠ACB = 120°.

Calculate

(i) the length CP
(ii) the length CB
(iii) the length AB
(iv) the angle of elevation of the top of the tower from B.

6. The waste-paper basket shown in Figure 7.68 has a top ABCD that is a square of side 30 cm and a base PQRS that is a square of side 20 cm.

The line joining the centres of the top and base is perpendicular to both and is 40 cm long.

Calculate

(i) the length PR
(ii) the length AC
(iii) the length AP.

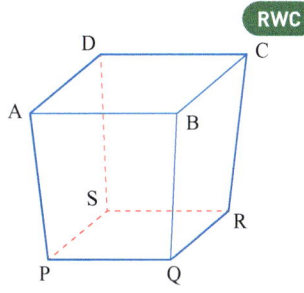

Figure 7.68

Lines and planes in three dimensions

7 In Egypt, pyramids were used as burial chambers for the Pharaohs.

The largest of these, shown in the diagram and built about 2500 BC for Cheops, is 146 m high and has a square base of side 231 m.

X is the centre of the base and VX = 146 m.

Calculate

Figure 7.69

(i) the angle between VA and ABCD
(ii) the length VA
(iii) the length VM, where M is the midpoint of AB
(iv) the angle between VAB and ABCD.

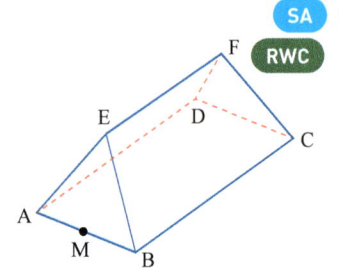

Figure 7.70

8 The tent shown in Figure 7.70 has a base that is 2.2 m wide and 3.6 m long.

The ends are isosceles triangles, inclined at an angle of 80° to the base.

∠AEB = ∠DFC = 70° and M is the midpoint of AB.

Calculate

(i) the length of EM
(ii) the height of EF above the base
(iii) the length of EF.

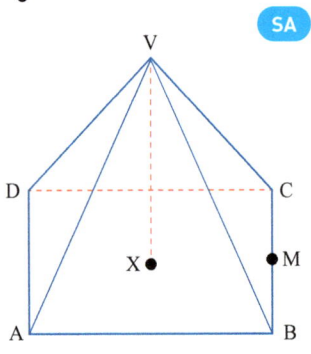

Figure 7.71

9 The right pyramid VABCD has rectangular base ABCD.

The vertex, V, is directly above the centre, X, of the base.

M is the midpoint of BC.

AB = 12 metres, BC = 9 metres and VA = 18 metres.

Work out

(i) the length AC
(ii) the length VX
(iii) the angle between VA and ABCD
(iv) the angle between VBC and ABCD.

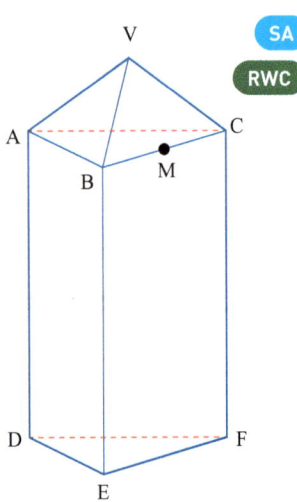

Figure 7.72

10 A new perfume is to be packaged in a box that is in the shape of a regular tetrahedron VABC of side 6 cm standing on a triangular prism ABCDEF as shown in the diagram.

The height of the prism is 12 cm.

M is the midpoint of BC.

Calculate

(i) the length AM
(ii) the length VM
(iii) the angle VAM
(iv) the total height of the box.

IR ⑪ The cuboid has a square base ABCD of side 6 cm and a height of 3 cm. M is the midpoint of EG.

(i) Calculate the length of BM.
(ii) Work out the area of triangle BEG.
(iii) Work out the angle between triangle BEG and the plane ABCD.

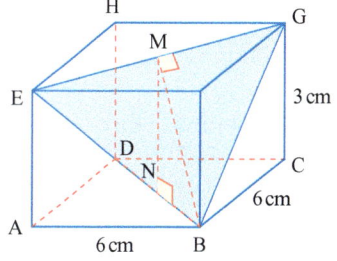

Figure 7.73

IR ⑫ The cube has sides of 12 cm and M is the midpoint of AC.

(i) Calculate the length of DM.
(ii) Work out the angle between the planes ABCD and ACH.
(iii) Calculate the area of the largest triangle that would fit inside this cube.
(iv) What is the area of the largest triangle that would fit inside a cube of side 20 cm? Give your answer in exact form.

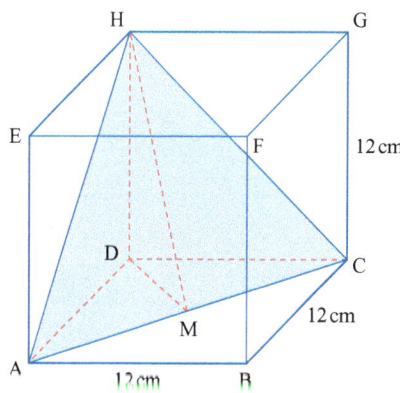

Figure 7.74

IR ⑬ A cuboid ABCDEFGH has edges of length 8 cm, 3 cm and 5 cm as shown.

Calculate the size of the smallest angle in triangle AEG.

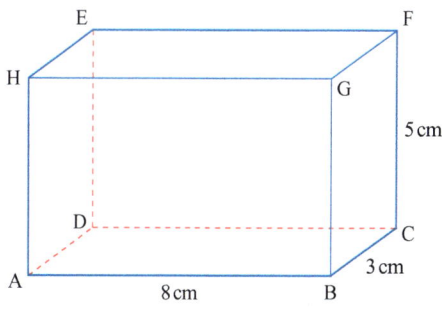

Figure 7.75

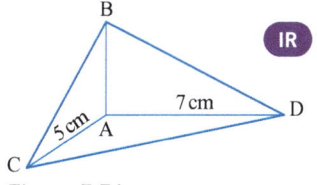

Figure 7.76

IR ⑭ A tetrahedron ABCD has lengths AC = 5 cm and AD = 7 cm.

Given that angle CAD = 130°, angle BCD = 65° and angle BDC = 55°, calculate the length of edge BC.

IR ⑮ Tetrahedron PQRS has lengths PQ = 8 cm, PS = 10 cm and QR = 5 cm, and angles QPS = 64° and QRS = 73°.

Calculate the size of angle QSR.

Learning outcomes

FUTURE USES
You will use the principles introduced here in the study of lines and planes in vector form at A-Level.

LEARNING OUTCOMES
Now you have finished this chapter, you should be able to
- calculate the area of a triangle given two sides and an included angle
- find an arc-length or a sector area by considering it as a fraction of a circle
- find an area of a segment by considering it as a triangle subtracted from a sector
- use the sine rule to calculate the size of an angle or a side-length
- use the cosine rule to calculate the size of an angle or side-length
- use the volume and surface area formulae for pyramids, cones and spheres
- draw a 2-D representation of a 3-D object
- calculate the angle between a line and a plane or the angle between two planes
- use Pythagoras' theorem to calculate lengths in three dimensions
- solve practical problems in three dimensions using the knowledge above.

REAL-WORLD CONTEXT
Angles between two lines, a line and a plane or two planes are of prime importance to architects and engineers in the design of buildings and machinery.

There are applications in navigation, for both ships and aircraft.

There are also applications in software engineering.

KEY POINTS

1. Area of a triangle $= \frac{1}{2}ab \sin C = \frac{1}{2}ac \sin B = \frac{1}{2}bc \sin A$

2. The length of an **arc** of a circle is calculated by considering it as a fraction of the circumference of the full circle.
$$\text{arc-length} = \frac{\alpha}{360} \times 2\pi r$$
Similarly, the area of a **sector** can be calculated as a fraction of the area of the full circle.
$$\text{sector area} = \frac{\alpha}{360} \times \pi r^2$$

3. Sine rule: $\frac{a}{\sin A} = \frac{b}{\sin B} = \frac{c}{\sin C}$ and $\frac{\sin A}{a} = \frac{\sin B}{b} = \frac{\sin C}{c}$

4. Cosine rule: $a^2 = b^2 + c^2 - 2bc \cos A$ and $\cos A = \frac{b^2 + c^2 - a^2}{2bc}$

5. When solving three-dimensional problems always draw a clear diagram where:
   - vertical lines are drawn vertically
   - east–west lines are drawn horizontally
   - north–south lines are drawn sloping
   - edges that are hidden are drawn as dotted lines.

6. In three dimensions, Pythagoras' theorem extends to
$$a^2 + b^2 + c^2 = d^2$$

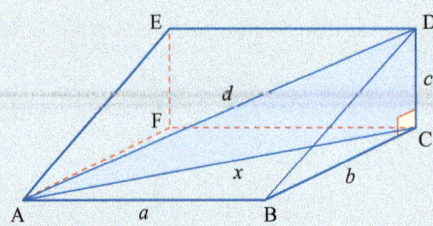

Figure 7.77

8 Calculus

> I do not know what I may appear to the world; but to myself I seem to have been only like a boy playing on the seashore, and diverting myself in now and then finding a smoother pebble or a prettier shell than ordinary, whilst the great ocean of truth lay all undiscovered before me.
>
> Isaac Newton

Prior knowledge

The formula 'gradient $= \dfrac{y_2 - y_1}{x_2 - x_1}$' was introduced in Chapter 3 for the gradient of a straight line joining the two points (x_1, y_1) and (x_2, y_2).

$m = \dfrac{y_2 - y_1}{x_2 - x_1}$ leads to a general equation of a straight line $(y - y_1) = m(x - x_1)$.

1 The gradient of a curve

In Figure 8.1 the curve has a zero gradient at A, a positive gradient at B and a negative gradient at C.

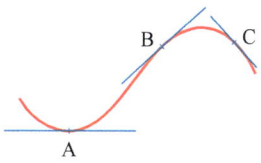

Figure 8.1

One way of finding these gradients is to draw the tangents and use two points on each one to calculate its gradient. This is time-consuming and the results depend on the accuracy of your drawing and measuring. If you know the equation of the curve, then *differentiation* provides another method of calculating the gradient.

2 Differentiation

Instead of trying to draw an accurate tangent, this method starts by calculating the gradients of chords $PQ_1, PQ_2, \ldots$. As the different positions of Q get closer to P, the values of the gradient of PQ get closer to the gradient of the tangent at P. For P at (3, 9), the first few positions of Q are shown in Figure 8.2.

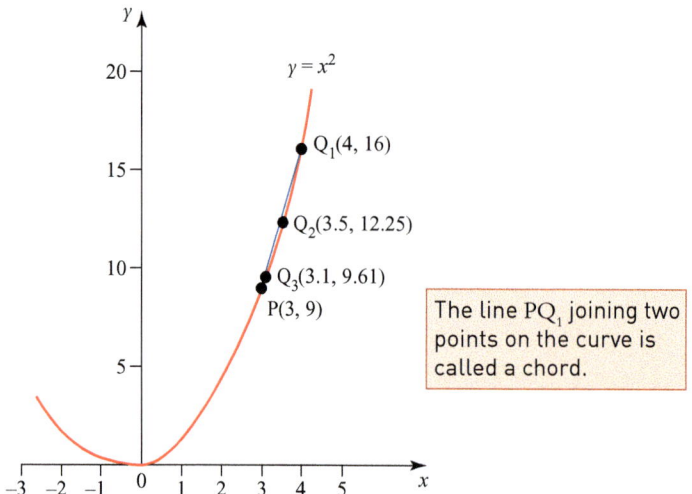

The line PQ_1 joining two points on the curve is called a chord.

Figure 8.2

chord	coordinates of Q	gradient of PQ
PQ_1	(4, 16)	$\dfrac{16 - 9}{4 - 3} = 7$
PQ_2	(3.5, 12.25)	$\dfrac{12.25 - 9}{3.5 - 3} = 6.5$
PQ_3	(3.1, 9.61)	$\dfrac{9.61 - 9}{3.1 - 3} = 6.1$
PQ_4	(3.01, 9.0601)	$\dfrac{9.0601 - 9}{3.01 - 3} = 6.01$
PQ_5	(3.001, 9.006 001)	$\dfrac{9.006\,001 - 9}{3.001 - 3} = 6.001$

ACTIVITY 8.1

Take points R_1 to R_5 on the curve $y = x^2$ with x-coordinates 2, 2.5, 2.9, 2.99, and 2.999 respectively and work out the gradients of the chords joining each of these points to P(3, 9).

In this process the gradient of the chord PQ gets closer and closer to that of the tangent, and hence the gradient of the curve at (3, 9).

Look at the sequence formed by the gradients of the chords.

7, 6.5, 6.1, 6.01, 6.001, …

It looks as though this sequence is converging to 6.

The table and the activity show that the gradient of the curve $y = x^2$ at (3, 9) seems to be 6 or about 6 but do not provide conclusive proof of its value. To do that you need to apply the method in more general terms.

Take the point P(3, 9) and another point Q close to (3, 9) on the curve $y = x^2$. Let the x-coordinate of Q be $(3 + h)$ where h is small. Since $y = x^2$ at all points on the curve, the y-coordinate of Q will be $(3 + h)^2$.

Figure 8.3 shows Q in a position where h is positive. Negative values of h would put Q to the left of P.

From Figure 8.3, the gradient of PQ is
$$\begin{aligned}\frac{(3+h)^2 - 9}{h} &= \frac{9 + 6h + h^2 - 9}{h} \\ &= \frac{6h + h^2}{h} \\ &= \frac{h(6 + h)}{h} \\ &= 6 + h.\end{aligned}$$

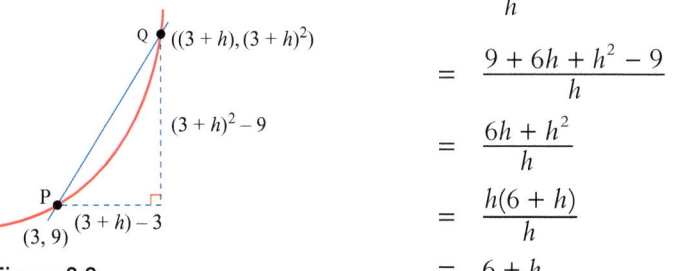

Figure 8.3

> **ACTIVITY 8.2**
> Using a similar method, work out the gradient of the tangent to the curve at
> (i) (2, 4)
> (ii) (−1, 1)
> (iii) (−3, 9).
> What do you notice?

For example, when $h = 0.001$, the gradient of PQ is 6.001 and when $h = -0.001$, the gradient of PQ is 5.999. The gradient of the tangent at P is between these two values. Similarly the gradient of the tangent at P would be between $6 - h$ and $6 + h$ for all small non-zero values of h.

For this to be true, the gradient of the tangent at (3, 9) must be *exactly* 6.

In this case, 6 was the *limit* of the gradient values, whether you approached P from the right or the left.

The gradient function

The work so far has involved calculating the gradient of the curve $y = x^2$ at just one particular point. It would be very tedious if you had to do this every time and so instead you can consider a general point (x, y) and then substitute the value(s) of x and/or y corresponding to the point(s) of interest. The gradient function is a measure of how the function is changing – often referred to as 'the rate of change of the function'.

> **Example 8.1**
> Calculate the gradient of the curve $y = x^3$ at the general point (x, y).
>
> ### Solution
> Let P have the general value x as its x-coordinate, so P is the point (x, x^3) (since it is on the curve $y = x^3$).
>
> Let the x-coordinate of Q be $(x + h)$ so Q is the point $((x + h), (x + h)^3)$.
>
>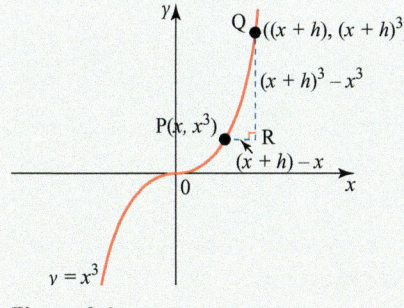
>
> **Figure 8.4**

Differentiation

The gradient of the chord PQ is given by

$$\frac{QR}{PR} = \frac{(x+h)^3 - x^3}{(x+h) - x}$$

$$= \frac{x^3 + 3x^2h + 3xh^2 + h^3 - x^3}{h}$$

$$= \frac{3x^2h + 3xh^2 + h^3}{h}$$

$$= \frac{h(3x^2 + 3xh + h^2)}{h}$$

$$= 3x^2 + 3xh + h^2$$

As Q gets closer to P, h takes smaller and smaller values and the gradient approaches the value of $3x^2$, which is the gradient of the tangent at P.

The gradient of the curve $y = x^3$ at the point (x, y) is equal to $3x^2$.

ACTIVITY 8.3

Use the method in Example 8.1 to prove that the gradient of the curve $y = x^4$ at the point (x, y) is equal to $4x^3$.

An alternative notation

So far, h has been used to denote the difference between the x-coordinates of our points P and Q, where Q is close to P.

h is sometimes replaced by δx. The Greek letter δ (delta) is shorthand for 'a small change in' and so δx represents a small change in x, δy a small change in y and so on.

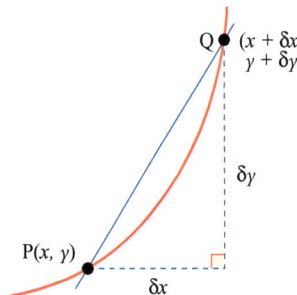

Figure 8.5

In Figure 8.5 the gradient of the chord PQ is $\dfrac{\delta y}{\delta x}$.

In the limit as δx tends towards 0, δx and δy both become infinitesimally small and the value obtained for $\dfrac{\delta y}{\delta x}$ approaches the gradient of the tangent at P.

> Read this as 'the limit as δx tends towards 0'.

$\displaystyle\lim_{\delta x \to 0} \dfrac{\delta y}{\delta x}$ is written as $\dfrac{dy}{dx}$.

Using this notation, you have a rule for differentiation.

$$y = x^n \Rightarrow \frac{dy}{dx} = nx^{n-1}$$

The gradient function, $\dfrac{dy}{dx}$, is sometimes called the *derivative* of y with respect to x and when you find it you have *differentiated* y with respect to x.

Because of the connection with gradient, $\dfrac{dy}{dx}$ is also referred to as the rate of change of y with respect to x.

Example 8.2

Given that $y = x^2 - 5x + 3$, find $\frac{dy}{dx}$ from first principles.

Solution

$$y = x^2 - 5x + 3$$
$$\therefore y + \delta y = (x + \delta x)^2 - 5(x + \delta x) + 3$$
$$\Rightarrow \delta y = (x + \delta x)^2 - 5(x + \delta x) + 3 - y$$
$$= x^2 + 2x\delta x + (\delta x)^2 - 5x - 5\delta x + 3 - (x^2 - 5x + 3)$$
$$= 2x\delta x + (\delta x)^2 - 5\delta x$$
$$= (2x - 5)\delta x + (\delta x)^2$$
$$\Rightarrow \frac{\delta y}{\delta x} = 2x - 5 + \delta x$$
$$\therefore \frac{dy}{dx} = \lim_{\delta x \to 0}(2x - 5 + \delta x)$$
$$= \frac{dy}{dx} = 2x - 5$$

Exercise 8A

① Find $\frac{dy}{dx}$ from first principles for each of the following equations.

(i) $y = 2x$
(ii) $y = 5x$
(iii) $y = 3x - 1$
(iv) $y = -4x + 7$

② Find $\frac{dy}{dx}$ from first principles for each of the following equations.

(i) $y = x^2$
(ii) $y = 3x^2$
(iii) $y = x^2 + 6$
(iv) $y = 4x^2 - 9$

③ Given that $y = x^2 + 4x + 9$, find $\frac{dy}{dx}$ from first principles.

④ Given that $y = 2x^2 - 7x - 6$, find $\frac{dy}{dx}$ from first principles.

⑤ Given that $y = 3 - 4x - 5x^2$, find $\frac{dy}{dx}$ from first principles.

Differentiation using standard results

Notation

An alternative way of expressing a function such as $y = 2x^2 + x - 3$ is to replace y by f(x) and write f(x) = $2x^2 + x - 3$. When discussing a function written in this form you just say 'f of x equals two x squared plus x minus three'. f'(x) is the notation used for the differential of f(x), so when y = f(x) you would write $\frac{dy}{dx}$ = f'(x). The derivative of f(x) is written as f'(x), pronounced as 'f dashed x'.

> ### ACTIVITY 8.4
>
> From earlier work you know that all lines of the form $y = x + c$ (where c can be positive, negative or zero) are parallel.
>
> Using any software at your disposal, sketch graphs of $y = x^2$, $y = x^2 + 5$ and $y = x^2 - 3$ on the same axes, setting your axes to $-4 < x < 4$ and $-5 < y < 25$.
>
> What do you notice?
>
> Repeat this for the graphs of $y = x^2 + 2x$, $y = x^2 + 2x + 5$ and $y = x^2 + 2x - 3$.

3 Differentiation using standard results

Finding the gradient from first principles establishes a formal basis for differentiation but in practice you would use the differentiation rule. This also includes the results obtained by differentiating (i.e. finding the gradient of) equations which represent straight lines.

The gradient of the line $y = x$ is 1.

The gradient of the line $y = c$ is 0, where c is a constant, since this line is parallel to the x-axis.

The rule can be extended further to include functions of the type $y = kx^n$ for any constant k, to give

$$y = kx^n \Rightarrow \frac{dy}{dx} = nkx^{n-1}.$$

You may find it helpful to remember the rule as

'multiply by the power of x and reduce the power by 1'.

Reflecting on Activity 8.4 and using this rule:

$y = x^2 + 2x \quad \Rightarrow \quad \frac{dy}{dx} = 2x + 2$

$y = x^2 + 2x + 5 \quad \Rightarrow \quad \frac{dy}{dx} = 2x + 2$

$y = x^2 + 2x - 3 \quad \Rightarrow \quad \frac{dy}{dx} = 2x + 2$

The three graphs have the same gradient function so are parallel.

Example 8.3

Write down the gradient function for each of the following functions.

(i) $y = x^7$ (ii) $y = 4x^3$ (iii) $y = 5x^2$ (iv) $y = x^{\frac{1}{3}}$

Solution

(i) $\dfrac{dy}{dx} = 7x^6$ 　　(ii) $\dfrac{dy}{dx} = 12x^2$

(iii) $\dfrac{dy}{dx} = 10x$ 　　(iv) $\dfrac{dy}{dx} = \dfrac{1}{3}x^{-\frac{2}{3}}$

Exactly the same rule, 'multiply by the power of x and reduce the power by 1' applies when the power is zero (i.e. y = a constant) or is negative. Remember that, for example, when you subtract 1 from 0, the answer is −1 and when you subtract 1 from −3 the answer is −4.

For $y = x^0$, the rule gives $\dfrac{dy}{dx} = 0 \times x^{-1}$ which is equal to 0.

Example 8.4

Work out the gradient function for each of the following functions.

(i) $y = x^{-3}$ 　　(ii) $y = 2x^{-4}$

(iii) $y = \dfrac{3}{x^2}$ 　　(iv) $y = \dfrac{3}{4x^2}$

(v) $y = \dfrac{6}{\sqrt{x}}$

Solution

(i) $\dfrac{dy}{dx} = -3x^{-4}$

(ii) $\dfrac{dy}{dx} = -8x^{-5}$

(iii) First write $y = \dfrac{3}{x^2}$ as $y = 3x^{-2} \Rightarrow \dfrac{dy}{dx} = -6x^{-3} = -\dfrac{6}{x^3}$

(iv) First write $y = \dfrac{3}{4x^2}$ as $y = \dfrac{3}{4}x^{-2} \Rightarrow \dfrac{dy}{dx} = \dfrac{3}{4}(-2x^{-3}) = -\dfrac{3}{2x^3}$

(v) First write $y = \dfrac{6}{\sqrt{x}}$ as $y = 6x^{-\frac{1}{2}} \Rightarrow \dfrac{dy}{dx} = -\dfrac{1}{2} \times 6x^{-\frac{3}{2}} = -\dfrac{3}{x^{\frac{3}{2}}}$

> You must leave the 4 in the denominator until you can simplify at the end.

Sums and differences of functions

Many of the functions you will meet are sums or differences of simpler ones. For example, the function $(4x^3 + 3x)$ is the sum of the functions $4x^3$ and $3x$. To differentiate a function such as this you differentiate each part separately and then add the results together.

Example 8.5

Differentiate $y = 4x^3 + 3x$.

Solution

$\dfrac{dy}{dx} = 12x^2 + 3$

Differentiation using standard results

Example 8.6

Differentiate $y = \dfrac{x^2}{2} - \dfrac{2}{3x^2}$.

Solution

Start by writing the expression in the form $y = \dfrac{1}{2}x^2 - \dfrac{2}{3}x^{-2}$

Differentiating, $\dfrac{dy}{dx} = \dfrac{1}{2}(2x) - \dfrac{2}{3}(-2x^{-3})$

$\qquad\qquad\quad = x + \dfrac{4}{3}x^{-3}$

$\qquad\qquad\quad = x + \dfrac{4}{3x^3}$

Example 8.7

Given that $y = 2x^3 - 3x + 4$, work out

(i) $\dfrac{dy}{dx}$

(ii) the gradient of the curve at the point (2, 14)

(iii) the rate of change of y with respect to x when $x = -3$.

Solution

(i) $\dfrac{dy}{dx} = 6x^2 - 3$

(ii) At (2, 14), $x = 2$

Substituting $x = 2$ in the expression for $\dfrac{dy}{dx}$ gives

$\dfrac{dy}{dx} = 6 \times (2)^2 - 3 = 21$.

(iii) $\dfrac{dy}{dx}$ is the rate of change of y with respect to x.

Substituting $x = -3$ in the expression for $\dfrac{dy}{dx}$ gives

$\dfrac{dy}{dx} = 6 \times (-3)^2 - 3$

$\qquad = 51$.

Example 8.8

Given that $y = 5x^2 - \dfrac{5}{x^2} + 2$, work out

(i) $\dfrac{dy}{dx}$

(ii) the gradient of the curve at the point (1, 2)

(iii) the rate of change of y with respect to x when $x = -1$.

Solution

(i) $y = 5x^2 - \dfrac{5}{x^2} + 2 \Rightarrow y = 5x^2 - 5x^{-2} + 2$

$\Rightarrow \dfrac{dy}{dx} = 10x - 5(-2)x^{-3}$

$= 10x + \dfrac{10}{x^3}$

(ii) At the point $(1, 2)$, $\dfrac{dy}{dx} = 10(1) + \dfrac{10}{1} = 20$.

(iii) $\dfrac{dy}{dx}$ is the rate of change of y with respect to x.

Substituting $x = -1$ in the expression for $\dfrac{dy}{dx}$ gives

$\dfrac{dy}{dx} = 10(-1) + \dfrac{10}{(-1)^3} = -10 + (-10) = -20$.

Exercise 8B

① Differentiate the following functions.

(i) $y = x^4$ (ii) $y = 2x^3$ (iii) $y = 5x^2$
(iv) $y = 7x^9$ (v) $y = -3x^6$ (vi) $y = 5$
(vii) $y = 10x$ (viii) $y = \dfrac{1}{4}x^3$ (ix) $y = 2\pi x$
(x) $y = \pi x^2$ (xi) $y = 4x^{\frac{3}{2}}$ (xii) $y = \sqrt[3]{8x}$

② Differentiate the following functions.

(i) $y = 2x^5 + 4x^2$ (ii) $y = 3x^4 + 8x$ (iii) $y = x^3 + 4$
(iv) $y = x - 5x^3$ (v) $y = 4x^3 + 2x$ (vi) $y = 2x + 6$
(vii) $y = 3x^5 + 2$ (viii) $y = 3x - x^{\frac{3}{5}}$ (ix) $y = 7x^{-3} - \sqrt[4]{x}$

③ Differentiate the following functions.

(i) $y = 3x^5 + 4x^4 - 3x^2 + 2$ (ii) $y = x^5 + 12x^3 + 3x$
(iii) $y = x^3 + 42x^2 - 5x + 24$ (iv) $y = 9x^{\frac{4}{3}} - 6x^{\frac{1}{3}}$

④ Write down the rate of change of the following functions with respect to y.

(i) $y = x^{-4}$ (ii) $y = 3x^{-2}$ (iii) $y = 3x^2 + 4x^{-1}$
(iv) $y = 2x^{-3} - 4$ (v) $y = x^2 + x^{-2}$ (vi) $y = 3x^{-2} + 2x^{-3}$
(vii) $y = 2x^{\frac{3}{2}} + 8x^{\frac{1}{2}} - 6x^{-\frac{1}{2}} + 4x^{-\frac{3}{2}}$

⑤ Differentiate the following functions.

(i) $y = 3x^2 + \dfrac{2}{x^3}$ (ii) $y = x^2 + \dfrac{1}{x^2}$ (iii) $y = 3x^3 + \dfrac{3}{x^3}$
(iv) $y = \dfrac{2}{x} - \dfrac{3}{x^2}$ (v) $y = \dfrac{1}{2x} - \dfrac{1}{3x^2}$ (vi) $y = \dfrac{2}{3x} - \dfrac{3}{4x^2}$
(vii) $y = 8x + \dfrac{7}{\sqrt{x}} - \dfrac{8}{\sqrt[3]{x}}$

⑥ A rectangle has length $6x$ and width $3x$.

The area of the rectangle is y.

(i) Write down y in terms of x.

(ii) Work out $\dfrac{dy}{dx}$.

Figure 8.6

Differentiation using standard results

⑦ When a stone is thrown into a lake, circular ripples appear centred on the point at which the stone entered the water and spreading outwards. After a time t seconds, the radius of the circle is r cm where $r = 10t^2$.

   (i) Work out the rate at which the radius is increasing (include the units).

   With time, the definition of the ripples becomes negligible so that after 8 seconds they cannot be seen by the human eye.

   (ii) What is the area of the largest ripple that you can see? Give your answer to the nearest 10 square metres.

⑧ An expanding sphere has radius $2x$.

   (i) Show that the volume, y, of the sphere is given by the formula $y = \frac{32}{3}\pi x^3$.

   (ii) Work out the rate of change of y with respect to x when $x = 2$.

Expressions that first need expanding or dividing

In this case you will need to manipulate the expression into a sum or difference before differentiating.

Example 8.9

Work out $\frac{dy}{dx}$ for each of the following equations.

(i) $y = x^3(x^2 - 4)$ (ii) $y = \frac{x^5 + x^2}{x}$

Solution

(i) Expand to give $y = x^5 - 4x^3$

$\Rightarrow \frac{dy}{dx} = 5x^4 - 12x^2$

(ii) Make into two fractions $y = \frac{x^5}{x} + \frac{x^2}{x}$

Cancel to give $y = x^4 + x$

$\Rightarrow \frac{dy}{dx} = 4x^3 + 1$

Exercise 8C

① Work out the gradient function for each of the following functions.

   (i) $y = x(x^2 + 2)$
   (ii) $y = 2x^2(3x - 4)$
   (iii) $y = (x + 3)(x + 2)$
   (iv) $y = (x + 5)(x + 2)$
   (v) $y = x^3(4 + x - x^2)$
   (vi) $y = (x + 2)(x - 5)$
   (vii) $y = \sqrt{x}(x^2 + 8x)$
   (viii) $y = (1 + \sqrt{x})(3 - \sqrt{x})$

② Work out an expression for the rate of change of y with respect to x for each of the following.

   (i) $y = \frac{x^5 + x^3}{4}$
   (ii) $y = \frac{x^7 + x^3}{x^2}$
   (iii) $y = \frac{4x^6 - 2x^2}{x^2}$
   (iv) $y = (3x + 1)(x - 2)$
   (v) $y = x^{\frac{1}{2}}(x^{\frac{3}{2}} + x^{\frac{1}{2}})$
   (vi) $y = x^{\frac{1}{2}}(x^{\frac{7}{2}} + x^{-\frac{1}{2}})$
   (vii) $y = \frac{x^{\frac{1}{2}} + x^{\frac{1}{3}}}{x^{\frac{1}{2}}}$
   (viii) $y = \frac{7x^{\frac{4}{3}} + 8x^{\frac{5}{3}}}{9x^{\frac{3}{2}}}$

③ (i) Simplify $\dfrac{3x^3 - 2x^2}{x}$.

(ii) Use your answer to (i) to differentiate $y = \dfrac{3x^3 - 2x^2}{x}$.

④ Work out the gradient of the curve $y = x^3(x - 2)$ at the point $(3, 27)$.

⑤ Work out the rate of change of y with respect to x for $\dfrac{6x^4 + 2x^5}{2x^3}$ when $x = -1$.

⑥ Work out the rate of change of y with respect to x for $y = x^{\frac{1}{3}}(x^{\frac{5}{3}} - x^{\frac{2}{3}})$ when $x = -3$.

⑦ Work out the gradient of the curve $y = \dfrac{3x^4 + x^2 - 5x}{x}$ at the point $(1, -1)$.

⑧ Work out the gradient of the curve $y = 3\sqrt{x} - \dfrac{3}{\sqrt{x}}$ at the point $(4, 4.5)$.

4 Tangents

Now that you know how to calculate the gradient of a curve at any point you can use this to work out the equation of the tangent at any particular point on the curve.

Example 8.10

(i) Work out the equation of the tangent to the curve $y = 3x^2 - 5x - 2$ at the point $(1, -4)$.

(ii) Sketch the curve and show the tangent on your sketch.

Solution

(i) First work out the gradient function $\dfrac{dy}{dx}$.

$$\dfrac{dy}{dx} = 6x - 5$$

Substitute $x = 1$ into this gradient function to calculate the gradient, m, of the tangent at $(1, -4)$.

$$m = 6 \times 1 - 5$$
$$= 1$$

The equation of the tangent is given by

$$y - y_1 = m(x - x_1)$$
$$y - (-4) = 1(x - 1) \quad \leftarrow \quad x_1 = 1, y_1 = -4 \text{ and } m = 1.$$
$$\Rightarrow \quad y = x - 5$$

(ii) $y = 3x^2 - 5x - 2$ is a ∪-shaped quadratic curve.

It crosses the x-axis when $3x^2 - 5x - 2 = 0$.

$\Rightarrow \quad (3x + 1)(x - 2) = 0$

$\Rightarrow \quad x = -\dfrac{1}{3}$ or $x = 2$

It crosses the y-axis when $y = -2$.

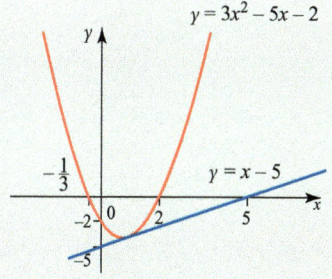

Figure 8.7

Tangents

Example 8.11

Figure 8.8 is a sketch of the curve $y = x^2 + \frac{1}{x}$ for $0 \leq x \leq 4$ where P is the point (2, 4.5).

Work out the equation of the tangent to the curve at P.

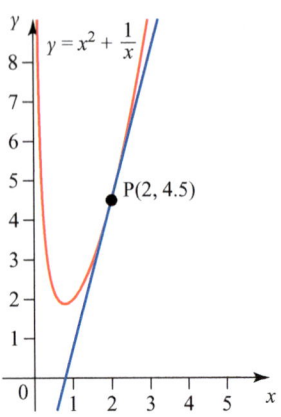

Figure 8.8

Solution

$$y = x^2 + \frac{1}{x} = x^2 + x^{-1} \Rightarrow \frac{dy}{dx} = 2x - x^{-2}$$

At (2, 4.5), $\frac{dy}{dx} = 4 - \frac{1}{4} = 3.75$ which is the gradient of the tangent.

Using $(y - y_1) = m(x - x_1)$ the equation of the tangent is

$$y - 4.5 = 3.75(x - 2)$$
$$\Rightarrow y - 4.5 = 3.75x - 7.5$$
$$\Rightarrow y = 3.75x - 3$$

Exercise 8D

① The sketch shows the graph of $y = 5x - x^2$.

The marked point, P, has coordinates (3, 6). Work out

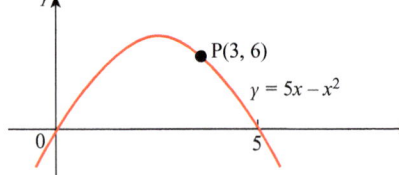

Figure 8.9

 (i) the gradient function $\frac{dy}{dx}$
 (ii) the gradient of the curve at P
 (iii) the equation of the tangent at P.

② The sketch shows the graph of $y = 3x^2 - x^3$.

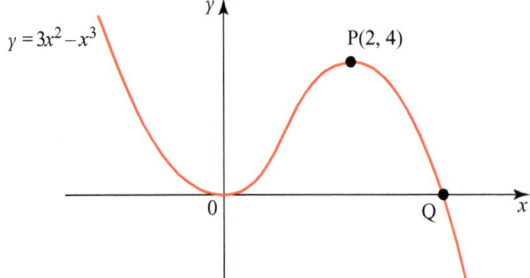

Figure 8.10

(i) The marked point, P, has coordinates (2, 4).
Work out the equation of the tangent at P.

(ii) The graph touches the x-axis at the origin O and crosses it at the point Q.
Work out the equation of the tangent at Q.

(iii) Without further calculation, state the equation of the tangent to the curve at O.

③ The sketch shows the graph of $y = x^5 - x^3$.

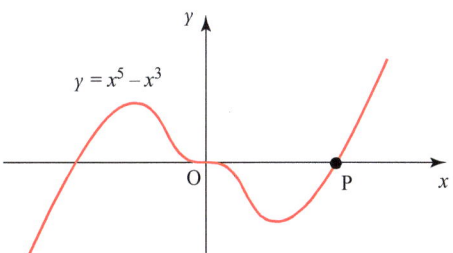

Figure 8.11

(i) Work out the coordinates of the point P where the curve crosses the positive x-axis.

(ii) Work out the equation of the tangent at P.

④ (i) Given that $y = x^3 - 3x^2 + 4x + 1$, work out the gradient function $\dfrac{dy}{dx}$.

(ii) The point P is on the curve $y = x^3 - 3x^2 + 4x + 1$ and its x-coordinate is 2.
Work out the equation of the tangent at P.

(iii) Work out the values of x for which the curve has a gradient of 13.

⑤ The sketch shows the graph of $y = x^3 - 9x^2 + 23x - 15$.

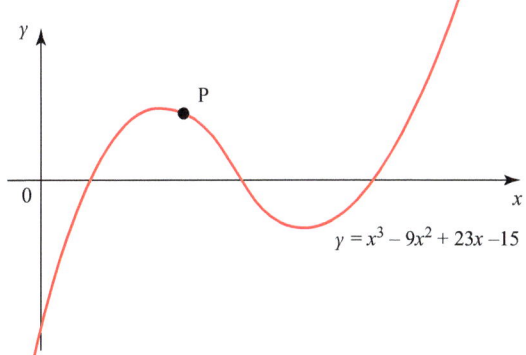

Figure 8.12

The point P marked on the curve has its x-coordinate equal to 2.

(i) Work out the equation of the tangent at P.

Q is a point on the curve where the tangent is parallel to the tangent at P.

(ii) Work out the equation of the tangent at Q.

Tangents

⑥ The point (2, −8) is on the curve $y = x^3 - px + q$.
   (i) Identify a relationship between p and q.
   The tangent to this curve at the point (2, −8) is parallel to the x-axis.
   (ii) Work out the value of p.
   (iii) Work out the coordinates of the other point where the tangent is parallel to the x-axis.

⑦ The sketch shows the graph of $y = x^2 - x - 1$.

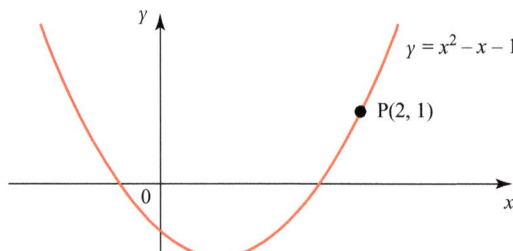

Figure 8.13

Work out the equation of the tangent at the point P.

⑧ A curve has the equation $y = (x - 3)(7 - x)$.
   Find the equation of the tangent at the point (6, 3).

⑨ A curve has the equation $y = 1.5x^3 - 3.5x^2 + 2x$.
   (i) Show that the curve passes through the points (0, 0) and (1, 0).
   (ii) Find the equations of the tangents at each of these points.

⑩ Figure 8.14 shows the curve with the equation $y = x^2 + \dfrac{2}{x}$ for $x > 0$.
   (i) Work out the gradient function $\dfrac{dy}{dx}$ and calculate the coordinates of the minimum point.
   (ii) State the equation of the tangent at that minimum point.
   (iii) Find the equation of the tangent at the point where $x = 2$.

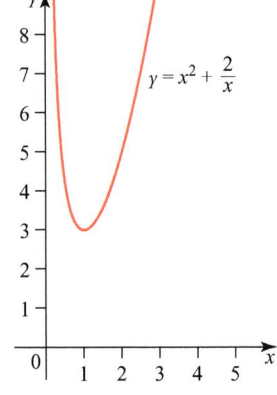

Figure 8.14

Discussion point

→ The curve in Figure 8.15 has equation $y = \dfrac{1}{x}$ for $x \neq 0$.

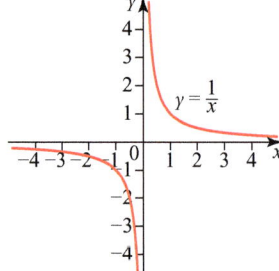

Figure 8.15

(i) Work out the equations of the tangents at the points (−1, −1) and (1, 1).
(ii) What do you notice about these lines?

5 Increasing and decreasing functions

A function $y = f(x)$ is

- increasing if $\dfrac{dy}{dx} > 0$
- decreasing if $\dfrac{dy}{dx} < 0$.

Some functions are increasing or decreasing over their whole domain.

For example, $y = 3 - 2x$ is a decreasing function for all real values of x because $\dfrac{dy}{dx} = -2$ which is < 0.

Other functions are increasing over parts of their domain and decreasing over others.

Example 8.12

Work out the values of x for which the function $y = x^2 - 4x + 1$ is an increasing function.

Solution

First work out $\dfrac{dy}{dx}$ $\qquad \dfrac{dy}{dx} = 2x - 4$

To be an increasing function $\dfrac{dy}{dx} > 0 \Rightarrow 2x - 4 > 0$

$$2x > 4$$
$$x > 2$$

Exercise 8E

1. Work out the values of x for which the following functions are increasing.
    - (i) $y = x^2 + 4$
    - (ii) $y = 2x - 3$
    - (iii) $y = x^2 + 2x - 5$
    - (iv) $y = x^2 - 3x$
    - (v) $y = 3x^2 + 4x + 7$
    - (vi) $y = (x + 6)(x - 2)$
    - (vii) $y = x^3 - 2x^2$
    - (viii) $y = x^3 + 6x^2 - 15x$
    - (ix) $y = x^3 - 3x^2 - 9x + 1$

2. Work out the values of x for which the following functions are decreasing.
    - (i) $y = 4x^2$
    - (ii) $y = x^2 - 6x + 2$
    - (iii) $y = x(x + 2)$
    - (iv) $y = 3 + 4x - x^2$
    - (v) $y = 12 - x$
    - (vi) $y = (2x + 1)^2$
    - (vii) $y = \dfrac{1}{3}x^3 + x^2$
    - (viii) $y = 2x^3 - 3x^2 - 72x$
    - (ix) $y = 27x - x^3$

3. Prove that $y = \dfrac{1}{3}x^3 + 2x^2 + 7x + 1$ is an increasing function for all values of x.

4. Prove that $y = x^3 - 6x^2 + 27x - 4$ is an increasing function for all values of x.

5. Work out the values of x for which $y = x^2 + \dfrac{2}{x}$ is an increasing function.

6. Prove that $y = 12 - 2x - x^3$ is a decreasing function for all values of x.

7. Prove that $y = \dfrac{1}{x}$ is a decreasing function for all $x \neq 0$.

The second derivative

⑧ Work out the values of x for which the following functions are
  (a) increasing   (b) decreasing.

  (i) $y = x + \dfrac{1}{x}$ (ii) $y = x - \dfrac{1}{x}$

  (iii) $y = x^2 + \dfrac{1}{x^2}$ (iv) $y = x^2 - \dfrac{1}{x^2}$

⑨ Air is being pumped into a spherical balloon at the rate of $1000\,\text{cm}^3\,\text{s}^{-1}$. Initially the balloon contains no air. (The formula for the volume of a sphere is $V = \dfrac{4}{3}\pi r^3$).

  (i) Calculate the volume V of the balloon after 10 seconds.
  (ii) Calculate the volume of the balloon after t seconds.
  (iii) State the value of $\dfrac{dV}{dt}$.
  (iv) Calculate the radius of the balloon after t seconds.

6 The second derivative

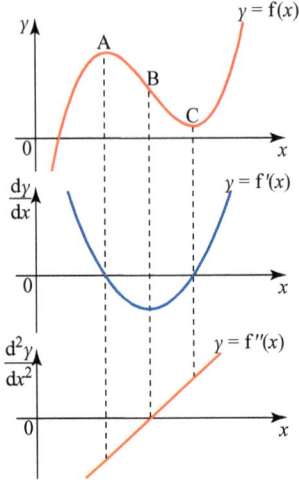

Figure 8.16

Figure 8.16 shows a sketch of a function $y = f(x)$ with a sketch of the corresponding gradient function, $\dfrac{dy}{dx} = f'(x)$ below it.

The third graph shows the gradient of the function $y = f'(x)$, denoted by $y = f''(x)$.

The gradient of any point on the curve of $\dfrac{dy}{dx}$ is found by differentiating $\dfrac{dy}{dx}$ and is given by $\dfrac{d}{dx}\left(\dfrac{dy}{dx}\right)$. This is written as $\dfrac{d^2y}{dx^2}$ or $y = f''(x)$ and is called the second derivative.

! $\dfrac{d^2y}{dx^2}$ is not the same as $\left(\dfrac{dy}{dx}\right)^2$.

Example 8.13

Given that $y = 2x^3 - 4x^2 + 3x - 1$, work out $\dfrac{d^2y}{dx^2}$.

Solution

$\dfrac{dy}{dx} = 6x^2 - 8x + 3$

$\dfrac{d^2y}{dx^2} = 12x - 8$

Example 8.14

A ball is thrown upwards with a speed of $20\,\text{m}\,\text{s}^{-1}$. Its height h m above the ground after a time of t seconds is given by $h = 1 + 20t - 5t^2$.

(i) Work out $\dfrac{dh}{dt}$ and say what this represents.

(ii) Calculate the maximum height reached by the ball and the time at which this height is reached.

(iii) Work out the rate of change of $\dfrac{dh}{dt}$, written as $\dfrac{d^2h}{dt^2}$, and say what this represents.

(iv) Sketch the graph of h against t.

Solution

(i) $\dfrac{dh}{dt} = 20 - 10t$

This represents the velocity of the stone.

(ii) The maximum height is reached when the ball is instantaneously at rest. This means that $\dfrac{dh}{dt} = 0$, giving $20 - 10t = 0$, so $t = 2$.

When $t = 2$, $h = 1 + 20(2) - 5(4) = 21$.

The maximum height is 21 m above the ground after a time of 2 seconds.

> When the velocity is positive the stone is moving upwards and when it is negative it is moving downwards. When it is zero it is stationary at the highest point.

(iii) $\dfrac{dh}{dt} = 20 - 10t \quad \Rightarrow \quad \dfrac{d^2h}{dt^2} = -10$

The rate of change of velocity is acceleration, and the positive direction is measured upwards, so this means that the acceleration of the ball is $-10\,\text{ms}^{-2}$ upwards, which is the same as saying that the ball is decelerating, i.e. slowing down, at a rate of $10\,\text{ms}^{-2}$ as it travels upwards. On its descent it will accelerate at a rate of $10\,\text{ms}^{-2}$ downwards.

(iv) $h = 1 + 20t - 5t^2$ is represented by a quadratic graph passing through $(0, 1)$ and having a maximum point at $(2, 21)$.

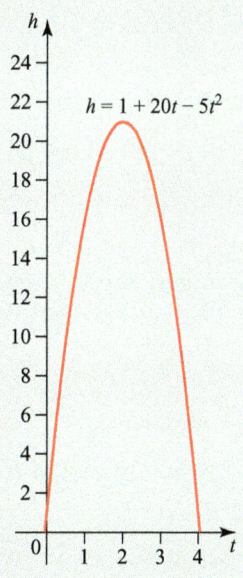

Figure 8.17

Exercise 8F

① Work out $\dfrac{dy}{dx}$ and $\dfrac{d^2y}{dx^2}$ for each of the following equations.

(i) $y = 3x^3 + 3x$ (ii) $y = x^5 - 25$

(iii) $y = 3x - 5x^4$

② Work out $\dfrac{dy}{dx}$ and $\dfrac{d^2y}{dx^2}$ for each of the following equations.

(i) $y = x^4 - 2x^2 + 5x - 4$ (ii) $y = 2x^3 + 3x - 4$

(iii) $y = x^3 - 2x^2 + 1$

Stationary points

③ Work out $\dfrac{dy}{dx}$ and $\dfrac{d^2y}{dx^2}$ for each of the following equations. Remember that when an expression involves brackets you need to multiply out before differentiating.

(i) $y = (2x - 1)(x + 2)$ 　　　　(ii) $y = (2x - 1)^2$

(iii) $y = (1 - 3x)(2x - 3)$

④ Work out $\dfrac{dy}{dx}$ and $\dfrac{d^2y}{dx^2}$ for each of the following equations.

(i) $y = 3(x - 2)(x^2 - 2x + 3)$ 　　　　(ii) $y = 2x^2(x - 1)^2$

(iii) $y = x^3(3x + 1)^2$

⑤ The sum of two numbers x and y is 13 and their product P is 40.

(i)   Write down an expression for y in terms of x.

(ii)   Write down an expression for P in terms of x.

(iii) Write down expressions for $\dfrac{dy}{dx}$ and $\dfrac{dP}{dx}$.

(iv)   Write down the rate of change of $\dfrac{dP}{dx}$.

⑥ For the curve $y = 3x^3 - 2x^2 - 6x - 4$

(i)   write down expressions for $\dfrac{dy}{dx}$ and $\dfrac{d^2y}{dx^2}$

(ii)   work out the gradient of the curve at the points $(-1, -7)$, $(1, -9)$ and $(2, 0)$

(iii) work out the rate of change of the gradient at each of these points.

⑦ A formula which you will meet in Mechanics or Physics is

$s = ut + \dfrac{1}{2}at^2$, where the letters in this case are $t =$ time, u is the initial velocity (which will be a constant, or zero if starting from rest), a is the acceleration (which must also be constant, for this formula) and s is the distance travelled. The only variables in the formula are s and t. Using this formula $\dfrac{ds}{dt}$ will give the velocity after a time t has elapsed.

(i)   Work out $\dfrac{ds}{dt}$ and hence the velocity after 12 seconds when the distance is measured in metres and time in seconds.

(ii) Work out $\dfrac{d^2s}{dt^2}$.

7 Stationary points

ACTIVITY 8.5

(i) Plot the graph of $y = x^4 - 3x^3 - x^2 + 3x$, taking values of x from -1.5 to $+3.5$ in steps of 0.5.
You will need your y-axis to go from -10 to $+20$.
Alternatively, if you have access to a graphics calculator or graphing software you could use that.

(ii) Describe the curve as x goes from -1.5 to 3.5.

A *stationary point* on a curve is one where the gradient is zero. This means that the tangents to the curve at these points are horizontal. Figure 8.18 shows a curve with two stationary points, A and B.

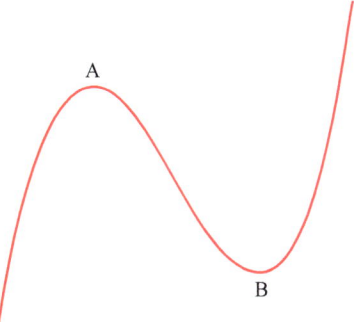

Figure 8.18

As the curve passes through the points A and B it changes direction completely. At A the gradient changes from positive to negative and at B from negative to positive. A is called a *maximum* point and B is a *minimum* point.

ACTIVITY 8.6

Figure 8.19 shows the graph of $y = \cos x$.
Describe the gradient of the curve, using the words 'positive', 'negative', 'zero', 'increasing' and 'decreasing', as x increases from 0° to 360°.

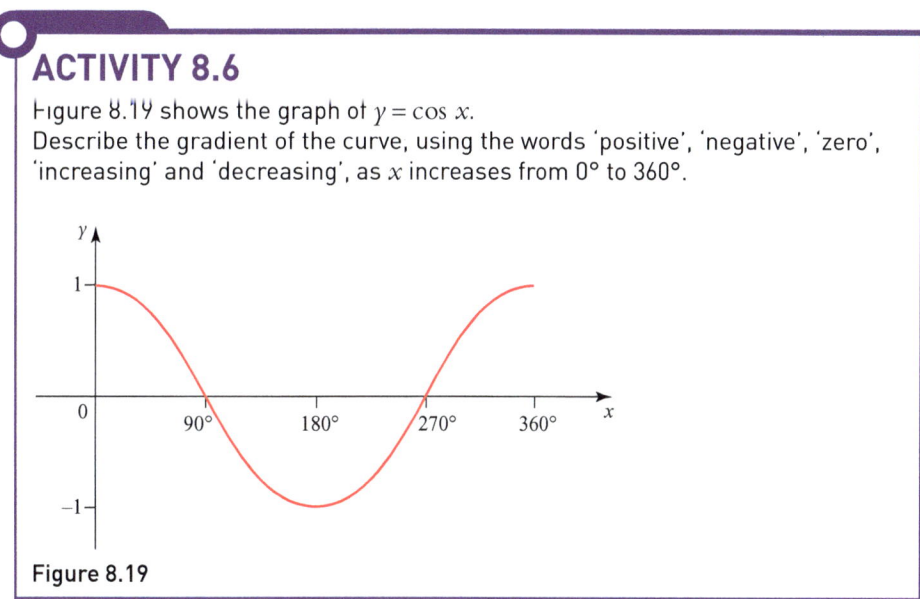

Figure 8.19

Maximum and minimum points

Figure 8.20 shows the graph of $y = 4x - x^2$. It has a maximum point at (2, 4).

You can see that

- at the maximum point the gradient $\dfrac{dy}{dx}$ is zero
- the gradient is positive to the left of the maximum and negative to the right of it.

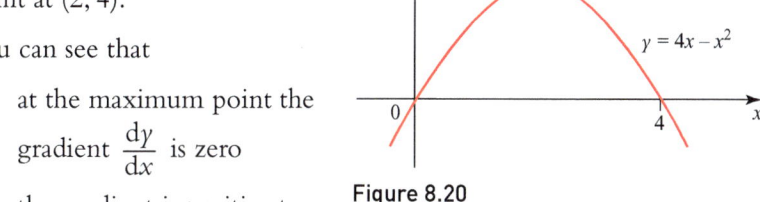

Figure 8.20

This is true for any maximum point (see Figure 8.21).

Stationary points

In the same way, for any minimum point (see Figure 8.22)

- the gradient is zero at the minimum
- the gradient goes from negative to zero to positive.

> You can see that the gradient function is decreasing $(+, 0, -)$ through a maximum point and increasing $(-, 0, +)$ through a minimum point.

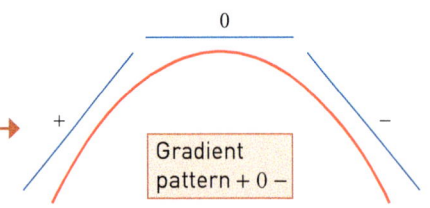

Gradient pattern $+ 0 -$

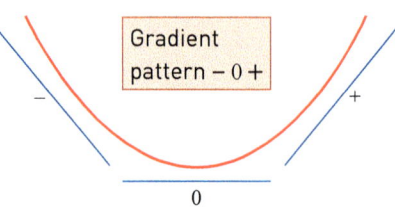

Gradient pattern $- 0 +$

Figure 8.21   Figure 8.22

Once you have found the position and type of any stationary points, you can use this information to sketch the curve.

Example 8.15

For the curve $y = x^3 - 12x + 3$

(i) work out $\frac{dy}{dx}$ and the values of x for which $\frac{dy}{dx} = 0$

(ii) classify the points on the curve with these x-values

(iii) work out the corresponding y-values

(iv) sketch the curve.

Solution

(i) $\frac{dy}{dx} = 3x^2 - 12$

When $\frac{dy}{dx} = 0$

$$3x^2 - 12 = 0$$
$$\Rightarrow 3(x^2 - 4) = 0$$
$$\Rightarrow 3(x + 2)(x - 2) = 0$$
$$\Rightarrow x = -2 \quad \text{or} \quad x = 2$$

(ii) For $x = -2$:

$x = -3 \Rightarrow \frac{dy}{dx} = 3(-3)^2 - 12 = +15$

$x = -1 \Rightarrow \frac{dy}{dx} = 3(-1)^2 - 12 = -9$

Gradient pattern is $+ 0 -$

$\Rightarrow$ maximum point when $x = -2$.

For $x = +2$:

$x = 1 \Rightarrow \frac{dy}{dx} = 3(1)^2 - 12 = -9$

$x = 3 \Rightarrow \frac{dy}{dx} = 3(3)^2 - 12 = +15$

Gradient pattern is $- 0 +$

$\Rightarrow$ minimum point when $x = +2$.

(iii) When $x = -2$, $y = (-2)^3 - 12(-2) + 3 = 19$.

When $x = +2$, $y = (2)^3 - 12(2) + 3 = -13$.

(iv) There is a maximum at $(-2, 19)$ and a minimum at $(2, -13)$.

The only other information you need to sketch the curve is the value of y when $x = 0$. This tells you where the curve crosses the y-axis.

When $x = 0$, $y = (0)^3 - 12(0) + 3 = 3$.

The graph of $y = x^3 - 12x + 3$ is shown in Figure 8.23.

> **Discussion point**
> → Why can you be confident about continuing the sketch of the curve beyond the x-values of the stationary points?

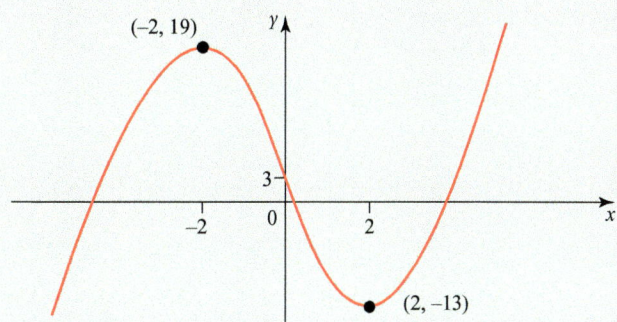

Figure 8.23

> **Discussion point**
> → In Example 8.15 you did not work out the coordinates of the points where the curve crosses the x-axis.
> (i) Why was this?
> (ii) Under what circumstances would you work out these points?

Example 8.16

Identify all the stationary points on the curve of $y = x^4 - 2x^3 + x^2 - 2$ and sketch the curve.

Solution

$\dfrac{dy}{dx} = 4x^3 - 6x^2 + 2x$

Stationary points occur when $\dfrac{dy}{dx} = 0$

$\Rightarrow \quad 2x(2x^2 - 3x + 1) = 0$

$\Rightarrow \quad 2x(2x - 1)(x - 1) = 0$

$\Rightarrow \quad x = 0$ or $x = 0.5$ or $x = 1$

You may find it helpful to summarise your working in a table. You can find the various signs, $+$ or $-$, by taking a test point in each interval, for example, $x = 0.25$ in the interval $0 < x < 0.5$.

Stationary points

	$x<0$	0	$0<x<0.5$	0.5	$0.5<x<1$	1	$x>1$
sign of $\dfrac{dy}{dx}$	−	0	+	0	−	0	+
stationary point		min		max		min	

When $x = 0$: $y = (0)^4 - 2(0)^3 + (0)^2 - 2 = -2$

When $x = 0.5$: $y = (0.5)^4 - 2(0.5)^3 + (0.5)^2 - 2 = -1.9375$

When $x = 1$: $y = (1)^4 - 2(1)^3 + (1)^2 - 2 = -2$

Therefore $(0.5, -1.9375)$ is a maximum stationary point and $(0, -2)$ and $(1, -2)$ are both minimum stationary points.

The graph of $y = x^4 - 2x^3 + x^2 - 2$ is shown in Figure 8.24.

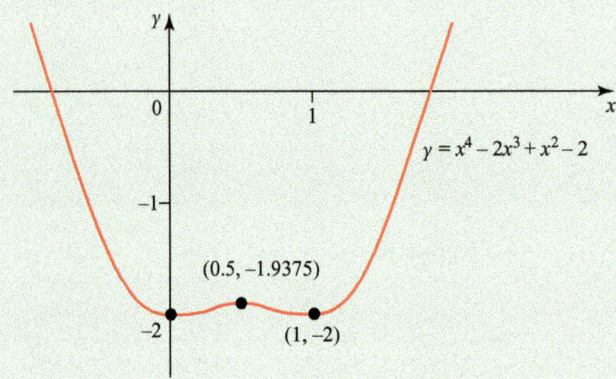

Figure 8.24

The method above, setting out a table of values, is rather tedious but there is an alternative way of identifying whether a stationary point is a maximum or a minimum using the second derivative. Recall that $\dfrac{d^2y}{dx^2}$ represents the rate of change of $\dfrac{dy}{dx}$, i.e. the rate of change of the gradient.

If $\dfrac{d^2y}{dx^2}$ is positive at a stationary point (i.e. where $\dfrac{dy}{dx} = 0$), then the gradient must go from negative to positive, in which case the turning point will be a minimum.

Conversely, if $\dfrac{d^2y}{dx^2}$ is negative at a stationary point, then the gradient must go from positive to negative which will indicate a maximum turning point.

Figures 8.21 and 8.22 illustrated that at a minimum/maximum the gradient function is increasing/decreasing. At a point where $\dfrac{dy}{dx} = 0$,

$\dfrac{d^2y}{dx^2} > 0$ gives a minimum and

$\dfrac{d^2y}{dx^2} < 0$ gives a maximum.

Note

If $\dfrac{d^2y}{dx^2} = 0$ at the stationary point, it is not possible to use this method and you will have to go back to the method of checking the gradient on each side of the stationary point.

Example 8.17

Given that $y = 2x^3 - 3x^2 - 12x + 4$

(i) work out $\dfrac{dy}{dx}$, and the values where $\dfrac{dy}{dx} = 0$

(ii) work out the coordinates of each of the stationary points

(iii) work out the value of $\dfrac{d^2y}{dx^2}$ at each of the stationary points and hence determine the nature of each one

(iv) sketch the curve.

Solution

(i) $\dfrac{dy}{dx} = 6x^2 - 6x - 12$

When $\dfrac{dy}{dx} = 0$, $\quad 6(x^2 - x - 2) = 0$

$\Rightarrow \quad 6(x - 2)(x + 1) = 0$

$\Rightarrow \quad x = 2$ or $x = -1$

(ii) When $x = 2$, $\quad y = 2(2)^3 - 3(2)^2 - 12(2) + 4$

$= -16$

When $x = -1$, $\quad y = 2(-1)^3 - 3(-1)^2 - 12(-1) + 4$

$= 11$

The stationary points are $(-1, 11)$ and $(2, -16)$.

(iii) $\dfrac{d^2y}{dx^2} = 12x - 6$

When $x = -1$, $\dfrac{d^2y}{dx^2} = -18 < 0$ so $(-1, 11)$ is a maximum point.

When $x = 2$, $\dfrac{d^2y}{dx^2} = +18 > 0$ so $(2, -16)$ is a minimum point.

(iv) The curve crosses the y-axis when $x = 0$, i.e. at the point $(0, 4)$. This information, together with the positions of the stationary points, is sufficient to enable you to sketch the curve.

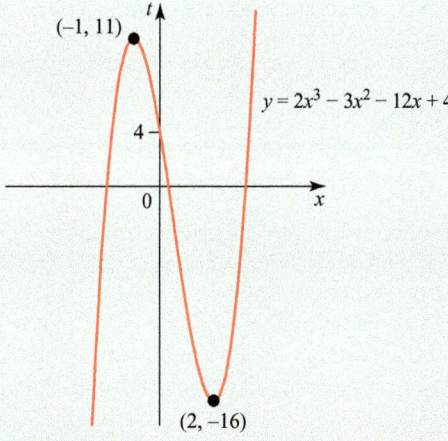

Figure 8.25

Stationary points

Exercise 8G

If you have access to a graphic calculator you will find it helpful to use it to check your answers.

1. For each of the curves given below
   - (a) work out $\frac{dy}{dx}$ and the value(s) of x for which $\frac{dy}{dx} = 0$
   - (b) work out the value(s) of $\frac{d^2y}{dx^2}$ at those points
   - (c) classify the point(s) on the curve with these x-values
   - (d) work out the corresponding y-value(s)
   - (e) sketch the curve.

   (i) $y = 1 + x - 2x^2$ (ii) $y = 12x + 3x^2 - 2x^3$

   (iii) $y = x^3 - 4x^2 + 9$ (iv) $y = x(x-1)^2$

   (v) $y = x^2(x-1)^2$ (vi) $y = x^3 - 48x$

   (vii) $y = x^3 + 6x^2 - 36x + 25$ (viii) $y = 2x^3 - 15x^2 + 24x + 8$

2. The graph of $y = px + qx^2$ passes through the point $(3, -15)$ and its gradient at that point is -14.
   - (i) Work out the values of p and q.
   - (ii) Calculate the maximum value of y and state the value of x at which it occurs.

3. (i) Identify the stationary points of the function $f(x) = x^2(3x^2 - 2x - 3)$ and distinguish between them.
   (ii) Sketch the curve $y = f(x)$.

4. The curve $y = ax^2 + bx + c$ crosses the y-axis at the point $(0, 2)$ and has a minimum point at $(3, 1)$.
   - (i) Work out the equation of the curve.
   - (ii) Check that the stationary point is a minimum.

FUTURE USES

- This work will be extended if you study Mathematics at a higher level.
- At A-Level you will learn additional formulae to deal with more complex algebraic products and quotients.
- There are also applications in other subjects, for example, Kinematics, Physics and Economics.

REAL-WORLD CONTEXT

Differentiation is used in the study of motion.
It is also the basis of differential equations which can be used to solve problems involving growth and decay.
Navier-Stokes equations, which are a particular form of differential equation, are vital to video-gaming and also help with the design of aircraft and cars, the study of blood flow, the design of power stations, the analysis of pollution and many other things.

LEARNING OUTCOMES

Now you have finished the chapter, you should be able to
- differentiate positive and negative powers of a variable such as x
- differentiate sums and differences of functions of x
- differentiate functions of x that first need expanding or dividing
- use differentiation to work out the gradient of a curve
- use this information to identify stationary points on a curve
- derive the equation of a tangent to a curve
- identify when a function is increasing and when it is decreasing
- calculate the position of any stationary points on the curve
- use the second derivative to determine the nature of any stationary points.

KEY POINTS

1. $y = kx^n \Rightarrow \dfrac{dy}{dx} = nkx^{n-1}$

   $y = c \Rightarrow \dfrac{dy}{dx} = 0$

   where n is a positive integer and k and c are constants.

2. $y = f(x) + g(x) \Rightarrow \dfrac{dy}{dx} = f'(x) + g'(x)$.

3. For the tangent and normal at (x_1, y_1)
   - the gradient of the tangent, m_1 = the value of $\dfrac{dy}{dx}$
   - the equation of the tangent is $y - y_1 = m_1(x - x_1)$.

4. A function $y = f(x)$ is increasing if $\dfrac{dy}{dx} > 0$.

   A function $y = f(x)$ is decreasing if $\dfrac{dy}{dx} < 0$.

5. The second derivative is obtained by differentiating $\dfrac{dy}{dx}$ and is denoted by $\dfrac{d^2y}{dx^2}$.

6. At a stationary point, $\dfrac{dy}{dx} = 0$.

   The nature of the stationary point can be determined by looking at the sign of the gradient just either side of it.

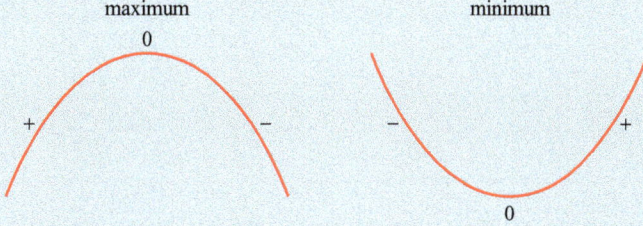

Figure 8.26

Key points

7 The sign of $\dfrac{d^2y}{dx^2}$ is an alternative way of determining the nature of the stationary point:

- If $\dfrac{d^2y}{dx^2} < 0$ at the stationary point, the point is a maximum.

- If $\dfrac{d^2y}{dx^2} > 0$ at the stationary point, the point is a minimum.

- If $\dfrac{d^2y}{dx^2} = 0$ at the stationary point, the result is inconclusive and you need to check the values of $\dfrac{dy}{dx}$ as in point 6.

9 Integration

Growth is painful. Change is painful. But nothing is as painful as staying stuck where you do not belong.

N.R. Narayana Murthy (1946–)

Discussion point

→ Suppose you know the gradient function, $\frac{dy}{dx}$, of a curve. What other information would you need to know to find the equation of the curve?

ACTIVITY 9.1

(i) Differentiate each of the following.
   (a) $y = x^3$ (b) $y = x^3 + 4$ (c) $y = x^3 - 7$
(ii) What do you notice?

The rule for integrating x^n where n is a positive integer

The equation $\frac{dy}{dx} = 3x^2$ is an example of a *differential equation*. All the equations in Activity 9.1 are solutions of this equation since they give an expression for y in terms of x. All that you can say at this point is that if $\frac{dy}{dx} = 3x^2$, then $y = x^3 + c$ where c is described as an *arbitrary constant*. An arbitrary constant can take any value, positive, negative or zero.

> The term 'constant of integration' is also used.

However, for any particular curve, the constant has a certain value which you will often have to determine.

1 The rule for integrating x^n where n is a positive integer

The rule for differentiation is usually given as

$$y = x^n \Rightarrow \frac{dy}{dx} = nx^{n-1}.$$

It can also be given as $y = x^{n+1} \Rightarrow \frac{dy}{dx} = (n+1)x^n$.

Dividing both right-hand sides by $(n+1)$ shows that this is the same as

$$y = \frac{1}{n+1} x^{n+1} \Rightarrow \frac{dy}{dx} = x^n.$$

Reversing this gives you the rule for integration.

$$\frac{dy}{dx} = x^n \Rightarrow y = \frac{1}{n+1} x^{n+1} + c$$

> Notice that for the statement to be true, the arbitrary constant c must be included.

Therefore integrating x^n with respect to x gives $\frac{x^{n+1}}{n+1} + c \ (n \neq -1)$.

You may find it helpful to remember the rule as

- add 1 to the power
- divide by the new power
- add a constant.

> **Note**
> As with differentiation, in this book if you are asked to integrate an expression f(x), take this to mean integrate with respect to x unless stated otherwise.

> **Discussion point**
> → Differentiating x gives 1, so integrating 1 gives $x + c$. How does this fit into the pattern above?

Example 9.1

Integrate the following.

(i) x^6 (ii) $5x^4$ (iii) 7

(iv) $2x^{-3}$ (v) $6x^{\frac{1}{2}}$

> ❗ Don't forget to include the arbitrary constant, c, until you have enough information to find a value for it.

Solution

(i) $\frac{x^7}{7} + c$ (ii) $5 \times \frac{x^5}{5} + c = x^5 + c$ (iii) $7x + c$

(iv) $\frac{2x^{-2}}{-2} + c = -x^{-2} + c$ (v) $\frac{6x^{\frac{3}{2}}}{\frac{3}{2}} + c = 4x^{\frac{3}{2}} + c$

$y = x^3 + c$ is called the *general solution* of the differential equation $\frac{dy}{dx} = 3x^2$ and the process of solving the equation in this way is called *integration*.

A solution such as this, containing an arbitrary constant, would give a family of curves, as in the diagram below. Each curve corresponds to a particular value of c.

> **Discussion point**
>
> → For how many equations is $\frac{dy}{dx} = 3x^2$?

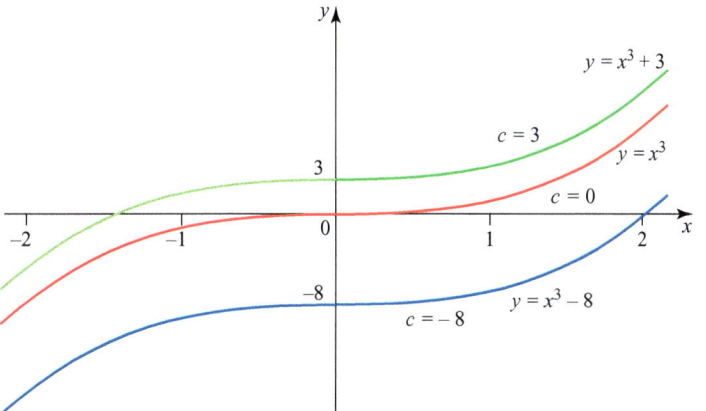

Figure 9.1

Suppose that in the previous activity you were also told that the solution curve passes through the point (1, 4). Substituting these coordinates in $y = x^3 + c$ gives

$$4 = 1^3 + c \quad \Rightarrow \quad c = 3$$

This example shows that if you know a point on a curve in the family, you can find the value of c, and so the equation of the curve. This is called the *particular solution*.

The particular solution in this case is $y = x^3 + 3$, which is one of the curves shown above.

Example 9.2

Given that $\frac{dy}{dx} = 6x^2 + 2x - 5$

(i) find the general solution of this differential equation

(ii) find the equation of the curve with this gradient function that passes through the point with coordinates (1, 7).

Solution

(i) By integration

$$y = 6 \times \frac{x^3}{3} + 2 \times \frac{x^2}{2} - 5x + c$$
$$= 2x^3 + x^2 - 5x + c, \text{ where } c \text{ is a constant.}$$

(ii) Since the graph passes through (1, 7),

$$7 = 2(1)^3 + 1^2 - 5 + c$$
$$\Rightarrow \quad c = 9$$
$$\Rightarrow \quad y = 2x^3 + x^2 - 5x + 9.$$

The rule for integrating x^n where n is a positive integer

Example 9.3

Find $f(x)$ given that $f'(x) = 2x + 4$ and $f(2) = -4$.

Solution

By integration
$$f(x) = \frac{2x^2}{2} + 4x + c$$
$$= x^2 + 4x + c, \text{ where } c \text{ is a constant.}$$

$f(2) = -4 \Rightarrow -4 = (2)^2 + 4(2) + c$

$\Rightarrow c = -16$

$\Rightarrow f(x) = x^2 + 4x - 16.$

Example 9.4

A curve passes through $(3, 5)$.

The gradient of the curve is given by $\frac{dy}{dx} = x^2 - 4$.

(i) Find y in terms of x.

(ii) Find the coordinates of any stationary points of the graph of y.

(iii) Sketch the curve.

Solution

(i) $\frac{dy}{dx} = x^2 - 4 \Rightarrow y = \frac{x^3}{3} - 4x + c$

When $x = 3$,
$$5 = 9 - 12 + c$$
$$\Rightarrow c = 8$$

So the equation of the curve is $y = \frac{x^3}{3} - 4x + 8$.

(ii) At all stationary points $\frac{dy}{dx} = 0$.

$\Rightarrow x^2 - 4 = 0$

$\Rightarrow (x + 2)(x - 2) = 0$

$\Rightarrow x = -2$ or $x = 2$

The stationary points are $(-2, 13\frac{1}{3})$ and $(2, 2\frac{2}{3})$.

(iii) The curve is a cubic with a positive x^3 term and two turning points, so it has this shape.

Figure 9.2

It crosses the y-axis when $x = 0, y = 8$.

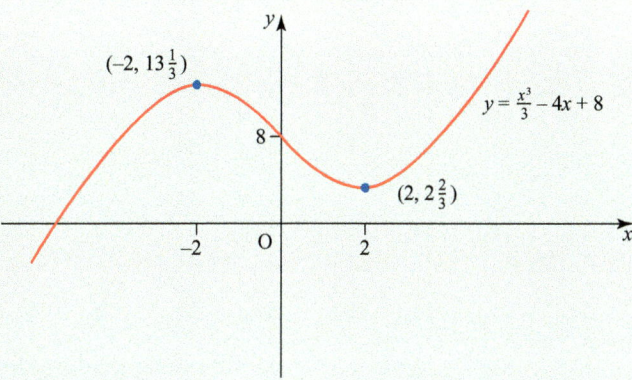

Figure 9.3

2 The integral notation

An alternative way of writing $\dfrac{dy}{dx} = 2x \implies y = x^2 + c$ is

$$\int 2x \, dx = x^2 + c.$$

> When integrating a function of x, it is essential that the function is followed by 'dx', which means 'with respect to x'.

Example 9.5

Find $\int \left(x^3 + \dfrac{1}{3x^2} - \sqrt{x} \right) dx$.

Solution

$$\int \left(x^3 + \dfrac{1}{3x^2} - \sqrt{x} \right) dx = \int \left(x^3 + \dfrac{x^{-2}}{3} - x^{\frac{1}{2}} \right) dx$$

$$= \dfrac{x^4}{4} + \dfrac{x^{-1}}{3 \times (-1)} - \dfrac{x^{\frac{3}{2}}}{\frac{3}{2}} + c$$

$$= \dfrac{x^4}{4} - \dfrac{1}{3x} - \dfrac{2x^{\frac{3}{2}}}{3} + c$$

> **Discussion point**
>
> → What would you need to do first before you could integrate $(2x + 1)(x - 4)$?

The integral notation

Exercise 9A

① For each of these gradient functions find $y = f(x)$.

(i) $\dfrac{dy}{dx} = 4x + 2$ (ii) $\dfrac{dy}{dx} = 6x^2 - 5x - 1$

(iii) $\dfrac{dy}{dx} = 3 - 5x^3$ (iv) $\dfrac{dy}{dx} = (x - 2)(3x + 2)$

(v) $f'(x) = 5x + 3$ (vi) $f'(x) = x^4 + 2x^3 - x + 8$

(vii) $f'(x) = (x - 4)(x^2 + 2)$ (viii) $f'(x) = (x - 7)^2$

(ix) $\dfrac{dy}{dx} = 7 + \sqrt[3]{x}$ (x) $f'(x) = \left(x^2 - \dfrac{1}{x}\right)^2$

② Find the following integrals.

(i) $\int 5x^3 \, dx$ (ii) $\int (2x - 3) \, dx$

(iii) $\int (3x^3 - 4x + 3) \, dx$ (iv) $\int (3 - x)^2 \, dx$

(v) $\int 4 \, dx$ (vi) $\int (2x + 1)(x - 3) \, dx$

(vii) $\int (x + 1)^2 \, dx$ (viii) $\int (2x - 1)^2 \, dx$

(ix) $\int \left(x + \dfrac{1}{\sqrt{x}}\right) dx$ (x) $\int \left(x + \dfrac{1}{x}\right)^2 dx$

③ For each of the following gradient functions, find the equation of the curve $y = f(x)$ that passes through the given point.

(i) $\dfrac{dy}{dx} = 2x - 3$; $(2, 4)$ (ii) $\dfrac{dy}{dx} = 4 + 3x^3$; $(4, -2)$

(iii) $\dfrac{dy}{dx} = 5x - 6$; $(-2, 4)$ (iv) $f'(x) = x^2 + 1$; $(-3, -3)$

(v) $f'(x) = (x + 1)(x - 2)$; $(6, -2)$ (vi) $f'(x) = (2x + 1)^2$; $(1, -1)$

(vii) $\dfrac{dy}{dx} = 5 - \sqrt{x}$; $(9, 2)$ (viii) $f'(x) = \left(2x^2 - \dfrac{3}{x}\right)^2$; $(1, 3)$

④ You are given that $\dfrac{dy}{dx} = 2x + 3$.

(i) Find the general solution of the differential equation.

(ii) Find the equation of the curve with gradient function $2x + 3$ and which passes through $(2, -1)$.

(iii) Hence show that $(-1, -13)$ also lies on the curve.

⑤ The curve C passes through the point $(3, 21)$ and its gradient at any point is given by $\dfrac{dy}{dx} = 3x^2 - 4x + 1$.

(i) Find the equation of the curve C.

(ii) Show that the point $(-2, -9)$ lies on the curve.

⑥ (i) Find the general solution of the differential equation $\dfrac{dy}{dx} = 4x - 1$.

(ii) Find the particular solution that passes through the point $(-1, 4)$.

(iii) Does this curve pass above, below or through the point $(2, 4)$?

⑦ The curve $y = f(x)$ passes through the point $(2, -4)$ and $f'(x) = 2 - 3x^2$.

Find the value of $f(-1)$.

⑧ A curve has stationary points at the points where $x = 0$ and $x = 2$.

(i) Explain why $\frac{dy}{dx} = x^2 - 2x$ is a possible expression for the gradient of the curve and give an alternative expression for $\frac{dy}{dx}$.

(ii) The curve passes through the point $(3, 2)$.

Taking $\frac{dy}{dx}$ as $x^2 - 2x$, find the equation of the curve.

3 Definite integrals

So far, the integrals you have met have all been *indefinite integrals*: they have finished with '$+ c$'. You may or may not have had some extra information to allow you to find a value for c.

Another form of integral, called a *definite integral*, is shown in Example 9.6.

Example 9.6

Find $\int_1^3 3x^2 \, dx$.

Prior knowledge

This uses exactly the same algebraic techniques as the previous 'indefinite' integrals, but the answer in this case has a numerical value.

Solution

$\int 3x^2 \, dx = x^3 + c$

To find the definite integral, you find the value of the integral when $x = 3$ and subtract the value when $x = 1$.

This gives $\int_1^3 3x^2 \, dx = \left[3^3 + c\right] - \left[1^3 + c\right] = 26$

> **Note**
>
> Notice how the c vanishes when you work out the expression above. When evaluating definite integrals, it is common practice to omit the c and write
>
> $\int_1^3 3x^2 \, dx = \left[x^3\right]_1^3 = \left[3^3\right] - \left[1^3\right] = 26.$

These numbers are called the *limits* of the integral; 3 is the *upper limit* and 1 is the *lower limit*.

The definite integral is defined as

$$\int_a^b f'(x) \, dx = [f(x)]_a^b = f(b) - f(a).$$

Example 9.7

Evaluate $\int_1^4 (x^2 + 3) \, dx$.

Discussion points

The word 'evaluate' is used to start Example 9.7.

→ What does that word mean?

→ Why is it appropriate to use it here?

Solution

$\int_1^4 (x^2 + 3) \, dx = \left[\frac{x^3}{3} + 3x\right]_1^4$

$= \left(\frac{4^3}{3} + 3 \times 4\right) - \left(\frac{1^3}{3} + 3 \times 1\right)$

$= 30$

Definite integrals

Example 9.8

Evaluate $\int_{-1}^{3}(x+1)(x-3)\,dx$.

Solution

$$\int_{-1}^{3}(x+1)(x-3)\,dx = \int_{-1}^{3}(x^2 - 2x - 3)\,dx$$

$$= \left[\frac{x^3}{3} - x^2 - 3x\right]_{-1}^{3}$$

$$= \left(\frac{3^3}{3} - 3^2 - 3 \times 3\right) - \left(\frac{(-1)^3}{3} - (-1)^2 - 3 \times (-1)\right)$$

$$= -10\frac{2}{3}$$

ACTIVITY 9.2

Evaluate

(i) $\int_{1}^{3} x^2\,dx$ and $\int_{3}^{1} x^2\,dx$

(ii) $\int_{-1}^{4}(x+3)\,dx$ and $\int_{4}^{-1}(x+3)\,dx$.

What do you notice?

(iii) What is the relationship between $\int_{a}^{b} f(x)\,dx$ and $\int_{b}^{a} f(x)\,dx$?

Exercise 9B

Evaluate the following definite integrals.

① $\int_{1}^{2} 3x^2\,dx$

② $\int_{1}^{4} 4x^3\,dx$

③ $\int_{-1}^{1} 6x^2\,dx$

④ $\int_{1}^{5} 4\,dx$

⑤ $\int_{2}^{4}(x^2 + 1)\,dx$

⑥ $\int_{-2}^{3}(2x + 5)\,dx$

⑦ $\int_{2}^{5}(4x^3 - 2x + 1)\,dx$

⑧ $\int_{5}^{6}(x^2 - 5)\,dx$

⑨ $\int_{1}^{3}(x^2 - 3x + 1)\,dx$

⑩ $\int_{-1}^{2}(x^2 + 3)\,dx$

⑪ $\int_{-4}^{-1}(16 - x^2)\,dx$

⑫ $\int_{1}^{3}(x+1)(3-x)\,dx$

⑬ $\int_{2}^{4}(3x(x+2))\,dx$

⑭ $\int_{-1}^{1}(x+1)(x-1)\,dx$

⑮ $\int_{-1}^{2}(x + 4x^2)\,dx$

⑯ $\int_{-1}^{1} x(x-1)(x+1)\,dx$

⑰ $\int_{-1}^{3}(x^3 + 2)\,dx$

⑱ $\int_{-3}^{1}(9 - x^2)\,dx$

⑲ $\int_{1}^{4}\left(\sqrt{x} + 2\right)\,dx$

⑳ $\int_{1}^{2}\left(\frac{1}{x^2} - 3\right)(1 + 5x^2)\,dx$

4 Areas between a curve and the x-axis

ACTIVITY 9.3

The diagram shows the line $y = 2x + 1$.
The shaded region is bounded by $y = 2x + 1$, the x-axis and the lines $x = 2$ and $x = 4$.

(i) Find the coordinates of the points A and B in the diagram.

(ii) Use the formula for the area of a trapezium to find the area of the shaded region.

(iii) Evaluate $\int_{2}^{4}(2x + 1)\,dx$ and confirm that your answer is the same as in part (ii).

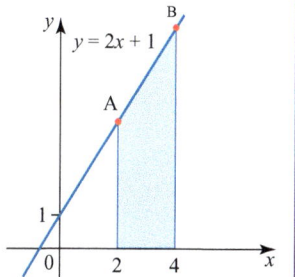

Figure 9.4

ACTIVITY 9.4

If possible, part (ii) of this activity should be done using a spreadsheet.

Figure 9.5 shows the graph of $y = x^2$ with the area between the curve, the x-axis and the lines $x = 0$ and $x = 3$ shaded.

(i) Calculate the area of the rectangles shaded in each of the two diagrams in Figure 9.6 and, for each one, say if you would expect this area to be larger or smaller than the shaded area between the curve and the x-axis.

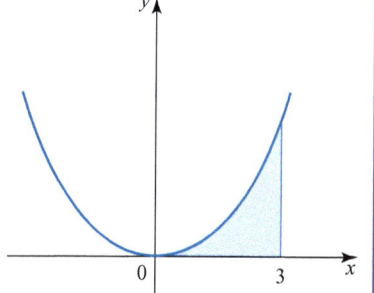

Figure 9.5

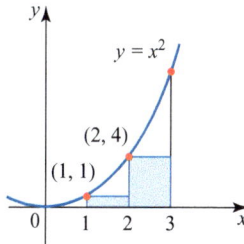 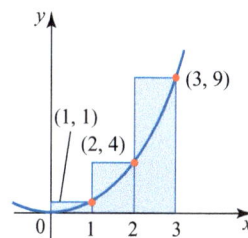

Figure 9.6

(ii) Now calculate the area of the rectangles when the width of the rectangles is reduced to

(a) 0.5

(b) 0.1.

(iii) Evaluate $\int_{0}^{3} x^2\,dx$.

(iv) What do you notice?

Areas between a curve and the x-axis

These activities illustrate a very important result: the area between a graph and the x-axis is given by a definite integral.

Note

The proof of this formula in general is not required at this level. It is an important result known as the fundamental theorem of calculus and the proof can be found in more advanced texts on pure mathematics.

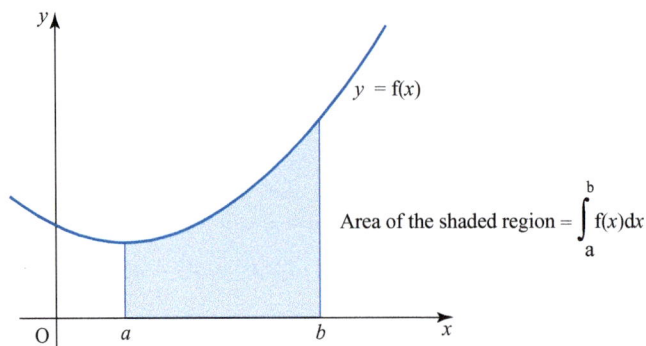

Area of the shaded region = $\int_a^b f(x)dx$

Figure 9.7

Example 9.9

The diagram below shows a sketch of the curve $y = 4 - x^2$.

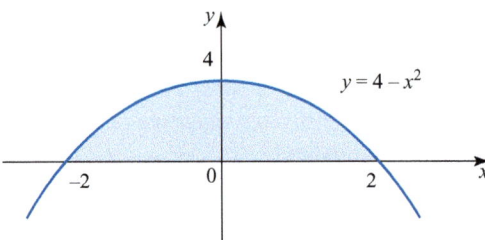

Figure 9.8

Find the area of the shaded region.

Solution

$$\text{Area} = \int_{-2}^{2} (4 - x^2) \, dx = \left[4x - \frac{x^3}{3} \right]_{-2}^{2}$$

$$= \left(4 \times 2 - \frac{2^3}{3}\right) - \left(4 \times (-2) - \frac{(-2)^3}{3}\right)$$

$$= 10\frac{2}{3} \text{ units}^2.$$

Exercise 9C

Find the area of each of the shaded regions.

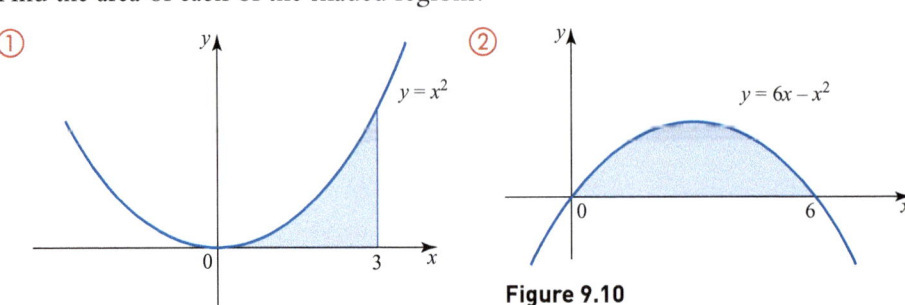

Figure 9.9

Figure 9.10

③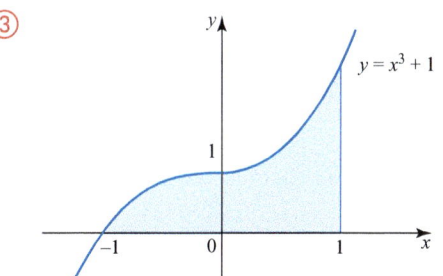

Figure 9.11

④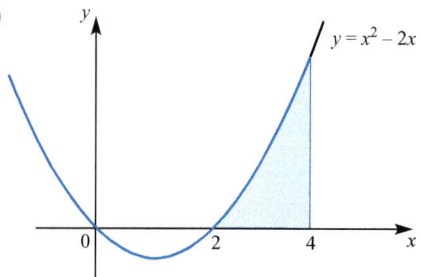

Figure 9.12

⑤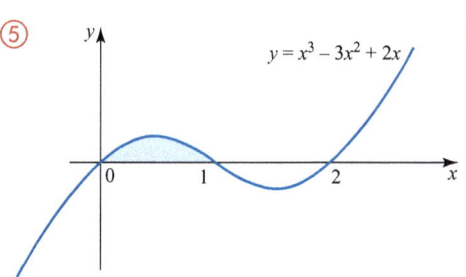

Figure 9.13

⑥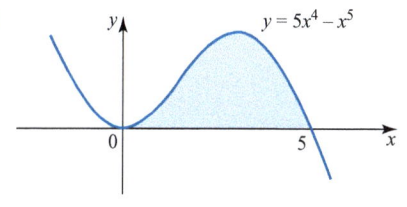

Figure 9.14

⑦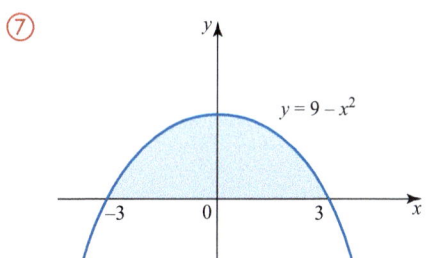

Figure 9.15

⑧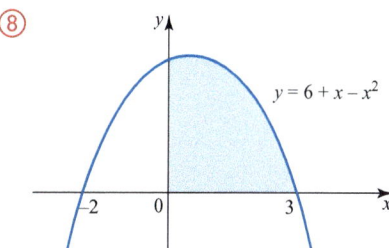

Figure 9.16

⑨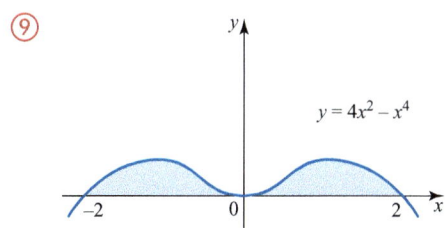

Figure 9.17

⑩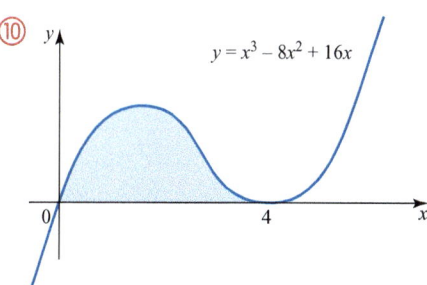

Figure 9.18

5 Areas below the x-axis

Example 9.10

The diagram below shows the line $y = x$ together with two regions marked A and B.

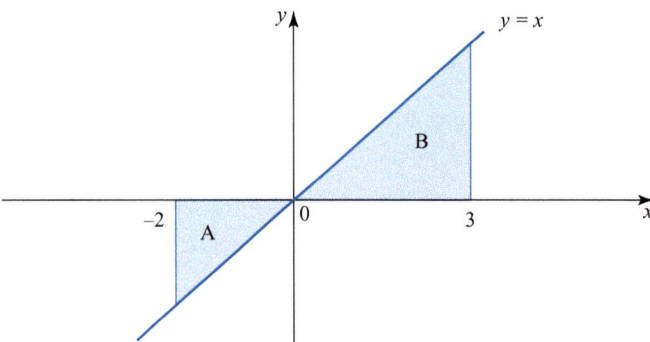

Figure 9.19

(i) Calculate the areas of A and B using the formula for the area of a triangle.

(ii) Evaluate $\int_{-2}^{0} x \, dx$ and $\int_{0}^{3} x \, dx$. What do you notice?

(iii) Evaluate $\int_{-2}^{3} x \, dx$. What do you notice?

Solution

(i) Area A $= \frac{1}{2} \times 2 \times 2 = 2$ units²

Area B $= \frac{1}{2} \times 3 \times 3 = 4.5$ units²

(ii) $\int_{-2}^{0} x \, dx = \left[\frac{x^2}{2}\right]_{-2}^{0} = 0 - (2)$

$= -2$

$\int_{0}^{3} x \, dx = \left[\frac{x^2}{2}\right]_{0}^{3} = 4.5 - 0$

$= 4.5$

Each area has the same numerical value as the integral but when the area is below the x-axis the integral is negative.

(iii) $\int_{-2}^{3} x \, dx = \left[\frac{x^2}{2}\right]_{-2}^{3} = 4.5 - (2)$

$= 2.5$

The areas above and below the x-axis have cancelled out.

! This example shows you how using integration gives a negative answer for the area of a region below the x-axis. In some contexts this will make sense and in others it won't, so you have to be careful.

Always draw a sketch graph when you are going to calculate areas.

Example 9.11

The diagram below shows a sketch of the curve $y = x(x-2)(x+2)$.

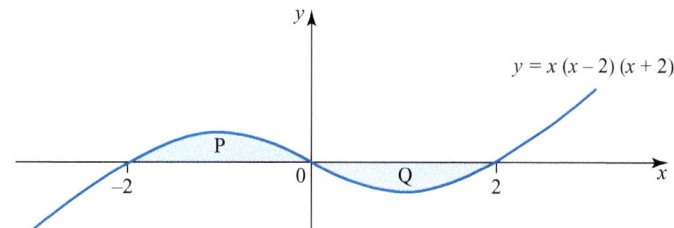

Figure 9.20

(i) Use integration to find the areas of each of the shaded regions P and Q.

(ii) Evaluate $\int_{-2}^{2} x(x-2)(x+2)\,dx$.

(iii) What do you notice?

Solution

(i) $\int_{-2}^{0} x(x-2)(x+2)\,dx = \int_{-2}^{0} (x^3 - 4x)\,dx$

$$= \left[\frac{x^4}{4} - 2x^2\right]_{-2}^{0}$$

$$= 0 - \left(\frac{(-2)^4}{4} - 2 \times (-2)^2\right)$$

$$= 4$$

So P has an area of 4 units².

$\int_{0}^{2} x(x-2)(x+2)\,dx = \int_{0}^{2} (x^3 - 4x)\,dx$

$$= \left[\frac{x^4}{4} - 2x^2\right]_{0}^{2}$$

$$= \left(\frac{2^4}{4} - 2 \times 2^2\right)$$

$$= -4$$

> The areas P and Q are the same since the curve has rotational symmetry about the origin.

So Q also has an area of 4 units².

(ii) $\int_{-2}^{2} x(x-2)(x+2)\,dx = \int_{-2}^{2} (x^3 - 4x)\,dx$

$$= \left[\frac{x^4}{4} - 2x^2\right]_{-2}^{2}$$

$$= \left(\frac{2^4}{4} - 2 \times 2^2\right) - \left(\frac{(-2)^4}{4} - 2 \times (-2)^2\right)$$

$$= 0$$

> Always draw a sketch graph when you are going to calculate areas. This will avoid any cancelling out of areas above and below the x-axis.

(iii) The areas of P and Q have 'cancelled out'.

Discussion point

→ In Example 9.11, would you say that the area between the curve and the x-axis is 0 units² or 8 units²?

Areas below the x-axis

Exercise 9D

① The sketch shows the curve $y = x^3 - x$.

Calculate the area of the shaded region.

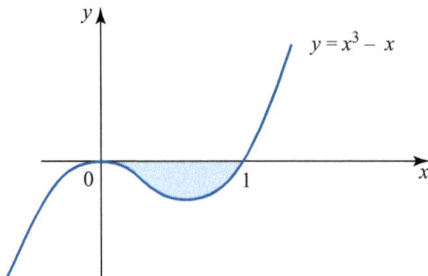

Figure 9.21

② The sketch shows the curve $y = x^3 - 4x^2 + 3x$.

(i) Calculate the area of the shaded regions P and Q.

(ii) State the area enclosed between the curve and the x-axis.

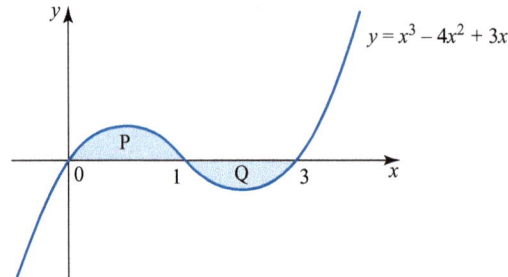

Figure 9.22

③ The sketch shows the curve $y = x^4 - 2x$.

(i) Find the coordinates of the point A.

(ii) Calculate the area of the shaded region.

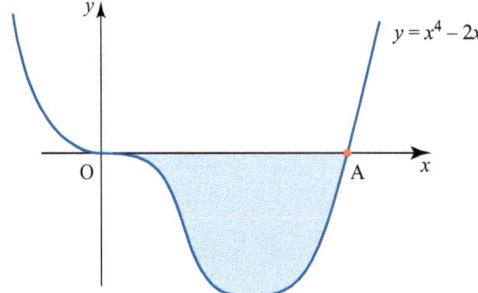

Figure 9.23

IR ④ The sketch shows the curve $y = x^3 + x^2 - 6x$.

Find the area between the curve and the x-axis.

> **Note**
>
> The WJEC Additional Mathematics specification only requires candidates to find the area of regions which are either wholly above the x-axis or wholly below it.

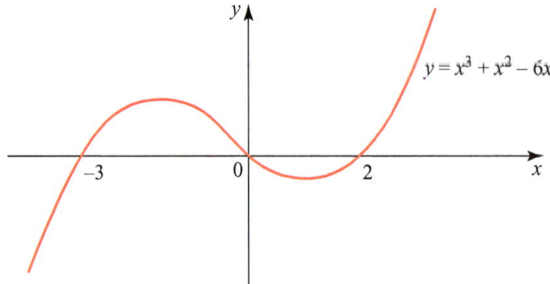

Figure 9.24

⑤ (i) Sketch the curve $y = x^2$ for $-3 \leq x \leq 3$.
  (ii) Shade the area bounded by the curve, the lines $x = -1$ and $x = 2$ and the x-axis.
  (iii) Find, by integration, the area of the region you have shaded.

⑥ (i) Sketch the curve $y = x^2 - 2x$ for $-1 \leq x \leq 3$.
  (ii) For what values of x does the curve lie below the x-axis?
  (iii) Find the area between the curve and the x-axis.

⑦ (i) Sketch the curve $y = x^3$ for $-3 \leq x \leq 3$.
  (ii) Shade the area between the curve, the x-axis and the line $x = 2$.
  (iii) Find, by integration, the area of the region you have shaded.
  (iv) Without any further calculation, state, with reasons, the value of
  $$\int_{-2}^{2} x^3 \, dx.$$

⑧ (i) Shade, on a suitable sketch, the region with an area given by
  $$\int_{-1}^{2} (x^2 + 1) \, dx.$$
  (ii) Evaluate this integral.

⑨ (i) Evaluate $\int_{1}^{4} (2x + 1) \, dx$.
  (ii) Interpret this integral on a sketch graph.

⑩ (i) Sketch the curve $y = (x - 1)(x - 3)$ and find the coordinates of the points where the curve crosses the x-axis.
  (ii) Calculate the area between the curve and the x-axis.
  (iii) Without any further calculation, explain why
  $$\int_{0}^{1} (x-1)(x-3) \, dx = \int_{3}^{4} (x-1)(x-3) \, dx.$$

⑪ The cross-section of a river bed has the equation $y = \dfrac{x^4}{40\,000}$, where x and y are measured in metres, as indicated in the diagram. The river is 40 m wide at surface level.

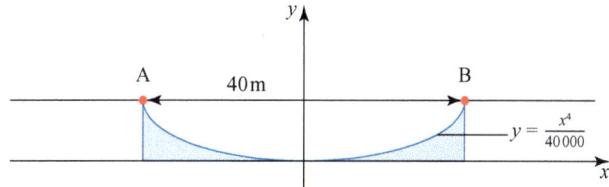

Figure 9.25

(i) Find the values of x at A and B.
(ii) Find the shaded area.
(iii) Hence find the area of the cross-section of the river bed.
(iv) The water flows at a speed of $1.6\,\text{m s}^{-1}$.
    How many cubic metres of water pass a particular point each hour?

Areas below the x-axis

12. The diagram shows a flower bed set into a rectangular garden. The rest of the garden is lawn. x and y are measured in metres.

    The equation of the curved edge of the flower bed is $y = \frac{1}{4}x^2 - 2x + 22$.

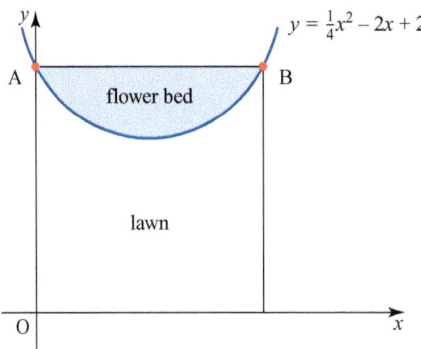

Figure 9.26

(i) Find the y-coordinate of the point A.

(ii) Hence find the x-coordinate of the point B.

(iii) Find the area of the lawn.

(iv) Hence find the area of the flower bed.

(v) The flower bed is to be covered with top soil to a depth of 30 cm. How many cubic metres of top soil are needed?

13. The child's slide shown in the diagram is bounded by parts of the curves $x = 0$, $y = 0.25$, $y = 0.125x^2 - 0.5x + 0.75$, $y = 2.25$ and $2x + y = 15.25$, as indicated. x and y are measured in metres.

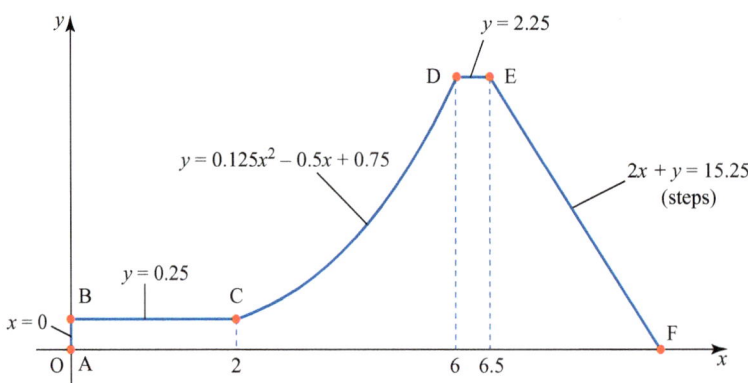

Figure 9.27

Children have been running underneath the slide and this is considered dangerous, so the council has decided to board up the space ABCDEF and cover it with a mural.

(i) Find the coordinates of point F.

(ii) Find the area of the board required to enclose one side of the slide.

6 The area between two curves

Example 9.12

Find the area enclosed between the line $y = 5 - x$ and the curve $y = x^2 - 3x + 5$.

Solution

You need to draw a sketch graph, but first you have to find where the curves intersect.

At the points of intersection

$x^2 - 3x + 5 = 5 - x$

$\Rightarrow \quad x^2 - 2x = 0$

$\Rightarrow \quad x(x - 2) = 0$

$\Rightarrow \quad x = 0$ or $x = 2$.

The curves intersect at $(0, 5)$ and $(2, 3)$.

$y = 5 - x$ is a line of gradient -1 passing through $(0, 5)$.

$y = x^2 - 3x + 5$ is a U-shaped quadratic also passing through $(0, 5)$.

The sketch is shown in Figure 9.28.

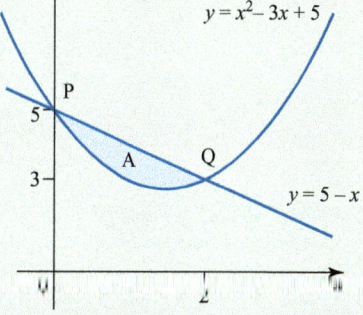

Figure 9.28

The shaded area can now be found in two ways.

Method 1

The area A can be treated as the difference between the two areas B and C.

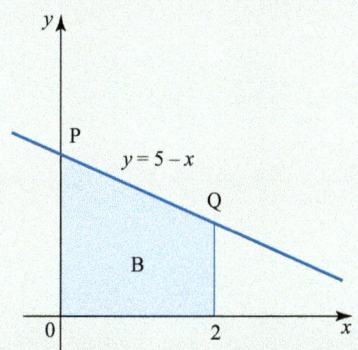 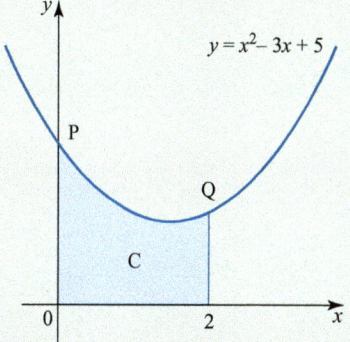

Figure 9.29

$A = B - C$

$= \int_0^2 (5 - x)\,dx - \int_0^2 (x^2 - 3x + 5)\,dx$

$= \left[5x - \dfrac{x^2}{2}\right]_0^2 - \left[\dfrac{x^3}{3} - \dfrac{3x^2}{2} + 5x\right]_0^2$

$= ((10 - 2) - 0) - \left(\left(\dfrac{8}{3} - 6 + 10\right) - 0\right)$

$= 1\dfrac{1}{3}$ units^2.

> **Note**
>
> The WJEC Additional Mathematics specification only requires candidates to find the area of regions bounded by a single curve, a horizontal line and/or vertical lines.

The area between two curves

Method 2

Notice how method 1 started by calculating

$$\text{Area} = \int_0^2 (\text{top curve})\,dx - \int_0^2 (\text{bottom curve})\,dx$$

These two integrals have the same limits, 0 and 2, so can be combined as

$$\int_0^2 (\text{top curve} - \text{bottom curve})\,dx$$

Discussion point

→ How does combining the integrals make your calculations easier?

$$= \int_0^2 ((5-x) - (x^2 - 3x + 5))\,dx$$

$$= \int_0^2 (2x - x^2)\,dx$$

$$= \left[x^2 - \frac{x^3}{3} \right]_0^2$$

$$= (4 - \tfrac{8}{3}) - 0 = 1\tfrac{1}{3} \text{ units}^2.$$

The big advantage of the second method is apparent when the area you require lies partly above and partly below the x-axis, as in the next example.

Example 9.13

The diagram below shows the curve $y = x^3 - 4x$ and the line $y = 8x + 16$, which is a tangent to the curve at the point $A(-2, 0)$.

The tangent meets the curve again at $B(4, 48)$.

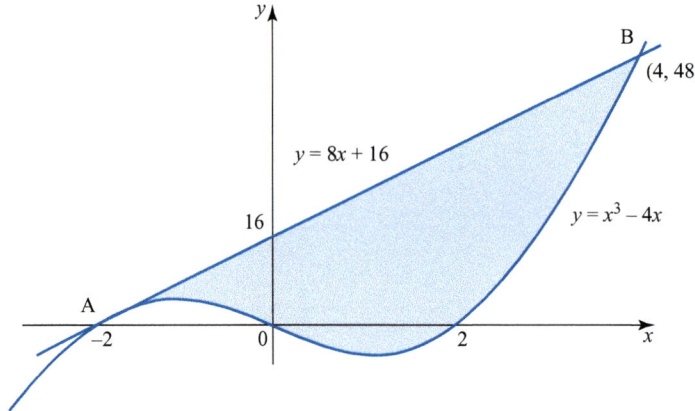

Figure 9.30

Find the area enclosed between the line and the curve.

Solution

$$\text{Area} = \int_{-2}^{4} (\text{top curve} - \text{bottom curve})\,dx$$

$$= \int_{-2}^{4} ((8x + 16) - (x^3 - 4x))\,dx$$

$$= \int_{-2}^{4} (12x + 16 - x^3)\,dx$$

Discussion point

→ Why was the second method so useful in this example?

$$= \left[6x^2 + 16x - \frac{x^4}{4} \right]_{-2}^{4}$$

$$= [(96 + 64 - 64) - (24 - 32 - 4)]$$

$$= 108 \text{ units}^2.$$

Exercise 9E

For each of questions 1–10

(i) sketch a graph of the given curves

(ii) find the x-coordinates of the points of intersection

(iii) find the area enclosed by the two curves.

① $y = 4 - x$; $y = (x - 1)(x - 4)$
② $y = x + 1$; $y = x^2 - 3x - 4$
③ $y = 1 - x$; $y = (x - 1)^2$
④ $y = 1 - x^2$; $y = x^2 - 1$
⑤ $y = x + 2$; $y = x^2 + x - 2$
⑥ $x + y = 9$; $y = x^2 - 2x + 3$
⑦ $y = x(x - 5)$; $y = x(10 - x)$
⑧ $y = 16 - x^2$; $y = x^2 - 5x + 13$
⑨ $y = x^2 - 16$; $y = 4x - x^2$
⑩ $y = x + 1$; $y = 5x - x^2 + 6$

⑪ A decorative mirror is bounded by the curves

$$y = \frac{x^2}{8} \quad \text{and} \quad y = 80 - \frac{x^2}{8}$$

and the lines

$x = 16$ and $x = -16$.

(i) Sketch the mirror.

(ii) Find the area of the mirror glass.

⑫ A sculpture is to be made up from a number of copper sheets, one of which is shown in the diagram.

The equations of the sides are

$$y = x(x + 3), \quad y = x - \frac{1}{4}x^2 \quad \text{and} \quad y = x^2 - 6x + 9.$$

All dimensions are in metres.

(i) Find which side has each equation.

(ii) Find the coordinates of A and B.

(iii) Calculate the area of the shape.

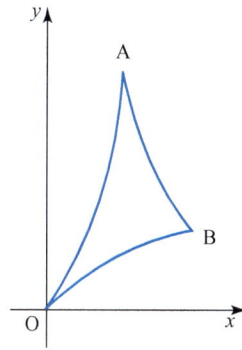

Figure 9.31

⑬ Some table mats are being designed in the shape of a flower, as shown in the diagram.

Each mat is made up of six equal sectors and all dimensions are in centimetres.

Line OC has equation $y = \sqrt{3}x$ and curve ABC has equation

$y = 16.2 - \dfrac{x^2}{12}$.

Line OB is a line of symmetry.

(i) Find the coordinates of the point C.

(ii) Find the area of OBC.

(iii) Hence find the total area of the table mat.

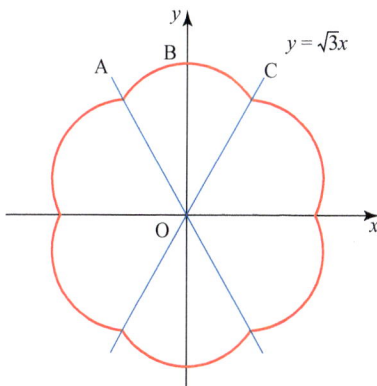

Figure 9.32

Learning outcomes

LEARNING OUTCOMES

Now you have finished this chapter, should be able to:

- integrate kx^n, where n is a positive integer or 0, and the sum of such functions
- understand integration as the reverse process of differentiation
- know what is meant by a definite and an indefinite integral
- evaluate definite integrals
- find the area between a curve, two ordinates and the x-axis
- find the area between two curves.

KEY POINTS

1. $\dfrac{dy}{dx} = x^n \Rightarrow y = \dfrac{x^{n+1}}{n+1} + c$

2. $\displaystyle\int_a^b x^n \, dx = \left[\dfrac{x^{n+1}}{n+1}\right]_a^b = \dfrac{b^{n+1} - a^{n+1}}{n+1}$

3. Area $A = \displaystyle\int_a^b y \, dx = \int_a^b f(x) \, dx$

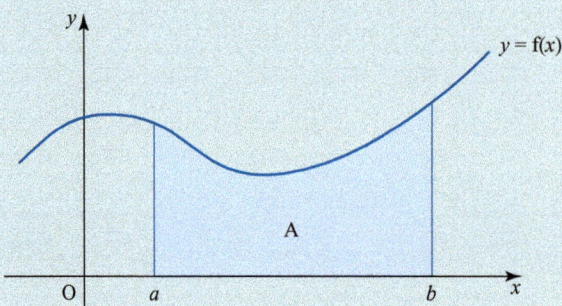

Figure 9.33

4. Areas below the x-axis give rise to negative values for the integral.

5. Area $B = \displaystyle\int_a^b (\text{top curve} - \text{bottom curve}) \, dx = \int_a^b (f(x) - g(x)) \, dx$

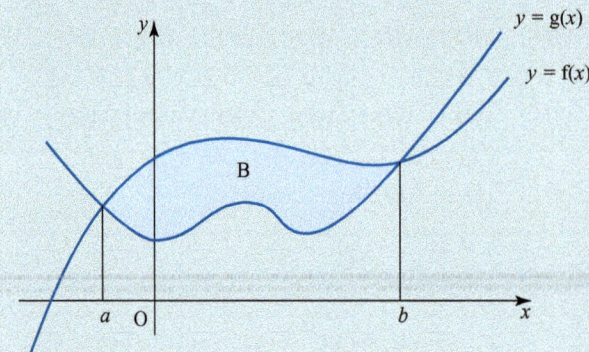

Figure 9.34

FUTURE USES

If you study mathematics at Advanced Level you will learn how to integrate trigonometric and exponential functions and much more.

Practice questions Paper 1

1. Find $\dfrac{dy}{dx}$ for each of the following.
   (a) $y = 3x^2 - 4x + 8$ [3 marks]
   (b) $y = \dfrac{5}{x^3}$ [1 mark]
   (c) $y = x^{\frac{3}{7}}$ [1 mark]

2. Showing all your working, simplify each of the following.
   (a) $\dfrac{3x^{\frac{3}{5}} \times 2x^{\frac{4}{5}}}{x^{\frac{3}{10}}}$ [2 marks]
   (b) $\dfrac{4y^{\frac{1}{6}} + y^{\frac{5}{6}}}{6y^{\frac{1}{6}}}$ [2 marks]

3. Without using a calculator, find the value of $(9^{\frac{1}{3}})^6$.
   Show all your working. [1 mark]

4. Prove that $\dfrac{x+1}{6} + \dfrac{3x}{4} - \dfrac{2x-3}{18} \equiv \dfrac{29x+12}{36}$. [4 marks]

5. (i) Factorise $16x^2 + 2x - 5$. [2 marks]
   (ii) **Hence** solve the equation $16x^2 + 2x - 5 = 0$. [2 marks]

6. (i) Write $x^2 + 8x + 13$ in the form $(x + a)^2 + b$. [2 marks]
   (ii) **Hence** write down the least value of $x^2 + 8x + 13$. [1 mark]
   (iii) What is the value of x when $x^2 + 8x + 13$ takes its least value? [1 mark]

7. Given that $y = x^2 + 5x$, find $\dfrac{dy}{dx}$ from first principles. [5 marks]

8. (a) Find the remainder when $x^3 + 9x^2 + 23x + 15$ is divided by $x - 2$. [2 marks]
   (b) (i) Show that $x + 1$ is a factor of $x^3 + 9x^2 + 23x + 15$. [2 marks]
   (ii) **Hence** factorise $x^3 + 9x^2 + 23x + 15$. [4 marks]

9. Find the coordinates and nature of each of the stationary points on the curve $y = x^3 - 27x + 8$.
   You must show all your working. [7 marks]

10. **Without using a calculator**, write $\dfrac{2}{5 - \sqrt{3}}$ in the form $\dfrac{a + \sqrt{b}}{c}$ where a, b and c are integers.
    You **must** show all your working. [3 marks]

11. Showing all your working, find the coordinates of the points of intersection of the curve $y = x^2 - 6x + 3$ and the line $y = x + 11$. [4 marks]

12 The curve $y = -x^2 + 6x - 5$ is shown here.

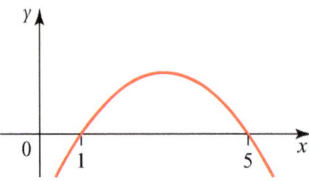

(a) Show that the points $(1, 0)$ and $(5, 0)$ lie on the curve
$y = -x^2 + 6x - 5$. **[2 marks]**

(b) Calculate the area of the region bounded by the curve
$y = -x^2 + 6x - 5$ and the x-axis.

You **must** show all your working. **[5 marks]**

13 (a) Find $\dfrac{d^2y}{dx^2}$ when $y = 3x^5 + 8x$. **[2 marks]**

(b) Find $\displaystyle\int \left(4x^3 + \dfrac{2}{x^2} - 9x\right) dx$. **[4 marks]**

(c) Showing all your working, evaluate $\displaystyle\int_1^4 (6x - 5)\, dx$. **[5 marks]**

14 Points A and B have coordinates $(1, 7)$ and $(-5, 11)$ respectively.

(a) Calculate the length of the line AB.
Write your answer as a simplified surd. **[3 marks]**

(b) Line L is perpendicular to AB and passes through the midpoint of AB.
Find an equation of L in the form $ax + by + c = 0$,
where a, b and c are integers. **[8 marks]**

15 Find the equation of the tangent to the curve $y = x^3 + x$
at the point where $x = 2$.
Express your answer in the form $y = mx + c$. **[6 marks]**

16 (a) Sketch the graph of $y = -4\cos x + 3$ for values of x from
$0°$ to $360°$.
Include the coordinates of any axis intercepts. **[3 marks]**

(b) State the maximum value and the minimum value
of $-4\cos x + 3$ for values of x from $0°$ to $360°$. **[1 mark]**

17 *You will be assessed on the quality of your written communication in this question.*
The edges of a square-based pyramid are all 10 cm.

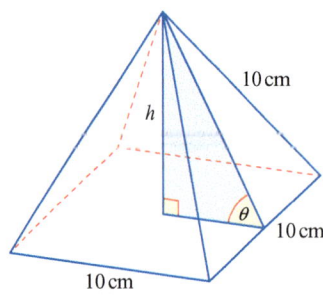

Diagram not drawn to scale.

(a) Calculate the perpendicular height, *h*, of the pyramid. [4 marks]

(b) Calculate the angle, θ, between each triangular face and the base. [3 marks]

⑱ The sides of the sector of a circle comprise two radii of length 10 cm and an arc of 12 cm as shown in the diagram.

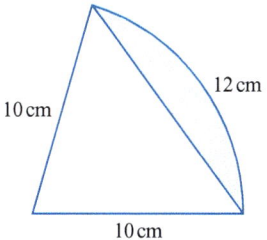

Find the area of the shaded segment. [5 marks]

Practice questions Paper 2

1. Find $\dfrac{dy}{dx}$ for each of the following.

    (a) $y = 5x^7 + x^3 - 6$ **[3 marks]**

    (b) $y = x^{-2}$ **[1 mark]**

    (c) $y = \sqrt{x}$ **[1 mark]**

2. Showing all your working, simplify each of the following.

    (a) $\dfrac{7y^{-\frac{1}{4}} \times 3y^{\frac{5}{4}}}{y^{\frac{1}{5}}}$ **[2 marks]**

    (b) $\dfrac{9x^{\frac{2}{5}} - 12x^{\frac{3}{5}}}{3x^{\frac{4}{5}}}$ **[2 marks]**

3. Without using a calculator, find the exact value of $64^{\frac{2}{3}} \times 36^{-\frac{1}{2}}$.
    Show all your working. **[2 marks]**

4. Bethan pays £23 for $(2x + 7)$ pens and £16 for $(3x - 1)$ pencils.
    Write an expression, in terms of x, for the total cost (in pounds) of 1 pen and 1 pencil.
    Leave your answer as a single fraction. **[5 marks]**

5. (i) Factorise $24x^2 + 5x - 1$. **[2 marks]**

    (ii) **Hence** solve the equation $24x^2 + 5x - 1 = 0$. **[2 marks]**

6. By completing the square, find the least value of $x^2 + 12x + 4$. **[3 marks]**

7. Given that $y = x^2 - 2x$, find $\dfrac{dy}{dx}$ from first principles. **[5 marks]**

8. (a) Find the remainder when $x^3 - 2x^2 - 13x - 10$ is divided by $x + 3$. **[2 marks]**

    (b) (i) Show that $x + 2$ is a factor of $x^3 - 2x^2 - 13x - 10$. **[2 marks]**

    (ii) **Hence** factorise $x^3 - 2x^2 - 13x - 10$. **[4 marks]**

9. Find the coordinates and nature of each of the stationary points on the curve $y = x^3 + 3x^2 - 7$.
    You must show all your working. **[7 marks]**

10. **Without using a calculator,** write $\dfrac{1}{6 + \sqrt{2}}$ in the form $\dfrac{a + \sqrt{b}}{c}$ where a, b and c are integers.
    You **must** show all your working. **[2 marks]**

11. Solve the simultaneous equations $y = 3x^2 + 5x - 8$ and $y = 2 - x$.
    Write your answers in the form $a + \dfrac{\sqrt{b}}{c}$ where a, b and c are integers.
    You **must** show all your working. **[6 marks]**

12. The curve $y = 8x - x^2$ is shown here.

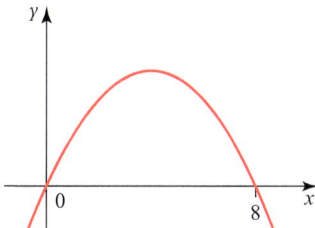

Calculate the area of the region bounded by the curve $y = 8x - x^2$ and the x-axis.

You **must** show all your working. [5 marks]

13. (a) Find $\dfrac{d^2y}{dx^2}$ when $y = 3x^{11}$. [2 marks]

(b) Given that $\dfrac{dy}{dx} = 6x^2 - 3$ and that $y = -6$ when $x = 1$, write y in terms of x. [3 marks]

14. (a) Find $\displaystyle\int \left(12x^5 - \dfrac{2}{x^3} + 2x\right) dx$. [4 marks]

(b) Showing all your working, evaluate $\displaystyle\int_0^3 (8x + 1)\,dx$. [5 marks]

15. Points A and B have coordinates $(-2, 9)$ and $(4, 1)$ respectively.

(a) Calculate the length of the line AB. [2 marks]

(b) Line L passes through points A and B.

Find an equation of L in the form $ax + by = c$, where a, b and c are integers. [6 marks]

16. Find the equation of the tangent to the curve $y = x^2 - 3x + 7$ at the point where $x = 5$.

Express your answer in the form $y = mx + c$. [6 marks]

17. (a) Sketch the graph of $y = 4\sin 2x$ for values of x from $0°$ to $180°$, including the coordinates of any axis intercepts.

Include the greatest and least y coordinates on the y-axis. [2 marks]

(b) Find all the solutions of the equation $4\sin 2x = 2$ for all values of x from $0°$ to $180°$. [3 marks]

18. ABCDEFGH is a cuboid with dimensions 3 cm, 4 cm and 12 cm.

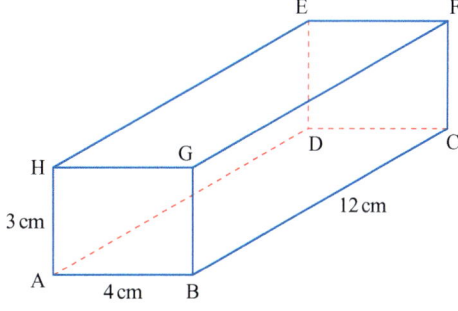

Diagram not drawn to scale.

Calculate the length of the diagonal AF. [4 marks]

19 *You will be assessed on the quality of your written communication in this question.*

ABCDE is a pentagon.

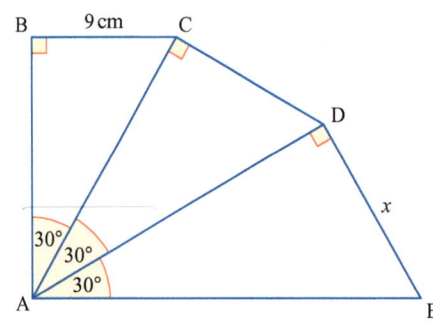

$A\hat{B}E = A\hat{C}D = A\hat{D}E = 90°$

$B\hat{A}C = C\hat{A}D = D\hat{A}E = 30°$

BC = 9 cm

Without using your calculator, and showing all your working, calculate the length of side DE, labelled x on the diagram. **[7 marks]**

Key words

Calculate	Work out the numerical value (often used after making a substitution)
Draw (a graph)	Draw axes on graph paper, plot points accurately and join with a straight line or smooth curve
Evaluate	Give a numerical value for your answer
Expand	Remove brackets
Expand and simplify	Remove brackets and collect like terms
Explain	Give reasons, either in words, or using mathematical symbols, or both
Expression	One or more terms, for example one side of a formula
Factorise	Write as a product
Give your answer in its simplest form	Cancel answers given as ratios or fractions or collect like terms
Hence	Use earlier work to deduce the result
Hence or otherwise	Using previous work to deduce the result is an option
Plot	Mark points (usually on graph paper) and join with a straight line or curve
Prove	Show all relevant steps (include explanations of facts used in geometrical proofs)
Show that	Show all relevant steps to reach a given result
Sketch (a graph)	Do not use graph paper. Draw axes and show the correct shape in each quadrant. Label appropriate points (e.g. intersection with axes, stationary points)
Verify	'Check out' a statement or result that you have been given
Work out the exact value of	Give the answer as an integer, fraction, recurring decimal, in terms of π, etc. or a surd
Write down	The answer should be obvious (no working is necessary)

Index

A

acceleration 171, 172
adjacent side 99–100
algebra
  algebra I 1–22
  algebra II 23–39
  algebra III 40–58
  algebra IV 59–83
  algebraic proof 80–3
  completing the square 35–9
  expanding brackets 12–14
  and factor theorem 74–80
  factorising 5–6, 23–8
  finding the equation of a line 48–53
  function notation 40–2
  graphs of functions 42–3
  graphs of linear functions 43–8
  graphs of quadratic functions 53–7
  indices with negative and fractional values 18–22
  linear equations 7–9, 34
  manipulating surds 14–18
  minimum value of a quadratic expression 37–9
  and number 10–11
  quadratic equations 53–7, 59–66, 74–5, 77
  rationalising the denominator 15–16
  rationalising denominators with two terms 17–18
  rearranging formulae 28–30
  remainder theorem 73–4
  simplifying algebraic fractions 30–3
  simplifying expressions 4–6, 14–17
  simultaneous equations 67–73
  solving linear equations involving fractions 34
algebraic proof 80–3
alternative notation 158–9
angles 133–5
  30°/45°/60° 104–7
  between a line and a plane 145, 154
  between two planes 145–9, 154
  and cosine graphs 108–9
  of depression 101
  and sine graphs 108–9
  and the sine rule 128–31
  and solution of trigonometric equations 111–14
  and tangent graphs 110
  trigonometric functions for 107–8
  and trigonometry in two dimensions 100–7
  use with the sine and cosine rule 135–8
arbitrary constants 182–3
arcs 126–8

area 6, 65
  5 areas below the x-axis 192–6
  between a curve and the x-axis 189–91
  between two curves 197–9
  of a circle 138
  of a triangle 106, 122–6, 138
  see also surface area
asymptotes 110, 119
axes
  areas between a curve and the x-axis 189–91
  drawing/plotting graphs 42
  sketching graphs 43

B

bearings 135–6
Bell, E.T. 59
brackets
  expansion 12–14, 29, 32, 35–6, 164–5
  and factorisation 23–6

C

calculators 75
calculus 155–80
  decreasing functions 169–70
  differentiation 156–60
  differentiation using standard results 160–5
  fundamental theorem 190
  gradient of a curve 155
  increasing functions 169–70
  real-world context 178
  second derivative 170–2
  stationary points 172–8, 184
  tangents 165–8
chords 156–9
circles
  arcs 126–8
  area 138
  circumference 28, 126, 138, 154
  radius 28, 124, 126, 128, 134, 139–41, 164, 170
  sectors 126–8
  segments 126–8
circumference 28, 126, 138, 154
coefficients 23–5
  equating 35–6, 55
  and the minimum value of a quadratic expression 37–8
  and solving quadratic equations by factorisation 60
completing the square 35–9
  solving quadratic equations by 60–1
cones
  right circular 139
  surface area 138, 139, 141
  volume 138–41

constant terms 24
constants
  equating 35–6, 55
  and solving quadratic equations by factorisation 60
coordinate geometry 84–97
  distance between two points 85
  equation of a straight line 90–3
  intersection of two lines 93–6
  midpoints of a line joining two points 86–9
  parallel lines 84–5
  perpendicular lines 84–5
cosine (cos) 99–101, 104–5
  definition 107–8
  $y = a \cos kx$ 114–19
cosine (cos) graphs 108–9, 112–14
cosine (cos) rule 132–5
  errors 133
  in three dimensions 150–3
  use with the sine rule 135–8
cubes 153
cubic equations 74–5
cuboids 6, 14
  and the angle between two planes 148–9
  and Pythagoras's theorem in three dimensions 149
  and the sine/cosine rule in three dimensions 150, 153
curves
  area between two curves 197–9
  areas between a curve and the x-axis 189–91
  gradient of a 155
  integration 181–7, 189–91, 193–9
  quadratic 62
  stationary points 173–8, 184
  tangents 165–8
  see also gradients; parabolas

D

definite integrals 187–8, 190
delta (δ) 158
denominators
  common 32
  rationalising 15–16
  rationalising with two terms 17–18
  and simplifying algebraic fractions 32
  and simplifying expressions containing square roots 15–17
derivatives, second 170–2
diagrams
  representations of three-dimensional objects 142–3
  true shape 143
difference of two squares 24

Index

differential equations 178, 182
   general solution 183
   particular solution 183
differentiation 156–60, 170, 178, 181–3
   rule for 182
   using standard results 160–5
distance between two points 85
division, expressions needing 164–5

E

Einstein, Albert 23, 84
elimination method 69–73
equations
   cubic 74–5
   differential 178, 182–3
   of a line 46–53
   linear 7–9, 34
   linear simultaneous 69–73
   Navier-Stokes 178
   quadratic 53–7, 59–66, 74–5, 77
   simultaneous 67–73, 93–5
   of a straight line 90–3
   trigonometric 110–14, 115–19
expansion 12–14, 29, 32
   and completing the square 35–6
   expressions that first need 164–5
expressions
   factorising 5–6, 23–4
   quadratic 23–4, 35–9
   simplifying 4–6, 14–17, 19–21
   that first need dividing/expanding 164–5

F

factor theorem 74–80
factorising 23–8, 29
   by inspection 23
   expressions 5–6, 23–4
   quadratic expressions 23–4
   solving quadratic equations by 59–60, 64
Faunce, Corporal John 40
flow charts, function notation 40
formulae, rearranging 28–30
fractions 2–3
   addition 3
   division 2, 3
   indices with negative and fractional values 18–22
   linear equations involving 34
   multiplication 3
   rationalising the denominator 15–16
   rationalising denominators with two terms 17–18
   simplifying algebraic fractions 30–3
   simplifying expressions containing square roots 15–16
   subtraction 2, 3
function notation 40–2
functions
   decreasing 169–70
   differences of 161–2
   graphs of linear 43–8
   graphs of 42–3
   increasing 169–70
   quadratic 53–7
   sums of 161–2
   see also specific functions
fundamental theorem of calculus 190

G

Galileo Galilei 1
general solution 183
geometry
   arcs 126–8
   area of a triangle 122–6
   cosine graphs 108–9, 111
   cosine rule 132–5
   geometry I 98–121
   geometry II 122–54
   lines and planes in three dimensions 143–53
   mensuration 138–41
   Pythagoras' theorem 98–9
   real-world context 120, 154
   sectors 126–8
   sine graphs 108–9
   sine rule 128–32
   solution of trigonometric equations 110–14
   tangent graphs 110
   three-dimensional problems 142–3
   trigonometric functions for angles of any size 107–8
   trigonometry in two dimensions 99–107
   using with the sine rule with the cosine rule 135–8
   $y = a \sin kx / y = a \cos kx / y = a \tan kx$ 114–19
   see also coordinate geometry
gradient functions 157–8, 160–1, 164–78, 181–7
gradients 155, 165–7, 170, 172, 197
   and chords 156–9
   and the equation of a straight line 91–2
   infinite 44
   and the intersection of two lines 94
   of a line 43–5, 48–51, 53
   midpoints of a line joining two points 86–8
   negative 44
   and parallel lines 84–5
   and perpendicular lines 84–8
   positive 44
   stationary points 173–4, 176
   zero 44
graphs
   cosine 108–9, 112–14
   drawing/plotting 42
   of functions 42–3
   of linear functions 43–8
   of quadratic functions 53–7
   and simultaneous equations 93–5
   sine 108–9, 112–14
   sketching 43
   trigonometric 115–19

H

hexagons 6, 125
Hipparchus 101
horizontal stretches 114–15
hypotenuse 99–100

I

indefinite integrals 187
index laws 18
indices, with negative and fractional values 18–22
integers, multiplying fractions by 34
integral notation 185–7
integrals 192, 198
   definite 187–8, 190
   indefinite 187
   limits of the 187
integration 181–200
   5 areas below the x-axis 192–6
   area between two curves 197–9
   area between a curve and the x-axis 189–91
   definite integrals 187–8
   integral notation 185–7
   rule for integrating x^n where n is a positive integer 182–5
intersection of two lines 93–6

K

Keynes, John Maynard 98

L

like terms, collection 4, 22, 32
limits of the integral 187
line of greatest slope 143
linear equations 7–9
   involving fractions 34
linear functions, graphs of 43–8
linear simultaneous equations 69–73
lines
   angle between a line and a plane 145
   asymptotes 110
   drawing/sketching a line given its equation 46–8
   equations of 45, 46–53
   gradients of 43–5, 48–51, 53
   intersection of two 93–6
   midpoints of a line joining two points 86–9
   parallel 84–5, 92, 144
   perpendicular 84–8, 91–2, 123, 145–6
   skew 144
   in three dimensions 143–53
   two lines in three dimensions 144

lines of symmetry 199
    and quadratic graphs 54–7
    and solving quadratic equations by the quadratic formula 62

M
manipulating surds 14–18
maximum points 173–8
mensuration 138–41
middle term, splitting the 24, 25, 26
midpoints of a line joining two points 86–9
minimum points 173–8
minimum value of a quadratic expression 37–9
Murthy, N. R. Narayana 181

N
Navier-Stokes equations 178
negative values, understanding and using indices with 18–22
Newton, Isaac 122, 155
Newton–Raphson method 75
notation 158–60
    alternative 158–9
    function 40–2
    integral 185–7
number 1–22
    and algebra 10–11
    number system 1–3
    numbers 1–3
numerators 15, 34

O
opposite side 99–100

P
parabolas 54, 58, 62
parallel lines 84–5, 92, 144
parallelograms 89, 134, 138
particular solution 183
pentagons 124
percentages 2, 3, 10–11
perimeters 6, 8
perpendicular lines 84–8, 91–2, 123, 145–6
planes
    angle between a line and a plane 145
    angle between two planes 145–9
    meeting in a line 145
    parallel 145
    in three dimensions 143–53
    two planes 145
polynomials
    and factor theorem 74–5, 79–80
    and remainder theorem 73–4
principal value 111–12
prisms 14, 138
products, negative 26

proof
    algebraic 80–3
    areas between a curve and the x-axis 190
    and the cosine rule 132
Ptolemy 101
pyramids
    and the angle between two planes 147
    and the sine/cosine rule in three dimensions 151, 152
    surface area 125, 140, 141
    volume 138, 140, 141
Pythagoras' theorem 85, 98–9, 101
    and the cosine rule 132
    in three dimensions 149
    in two dimensions 104

Q
quadratic curves 62
quadratic equations 59–66
    and factor theorem 74–5, 77
    formulation and manipulation 64–6
    solving by completing the square 60–1
    solving by drawing a graph 53–7
    solving by factorising 59–60, 64
    solving by the quadratic formula 61–3
quadratic expressions
    completing the square 35–9
    factorising 23–4
    minimum value of a 37–9
quadratic factors 78
quadratic formula, the 61–3
quadratic functions, graphs of 53–7
quadrilaterals
    and the cosine rule 134
    and the equation of a straight line 92
    and intersection of two lines 96
    and midpoints of a line joining two points 88–9
    and the sine rule 131
quartic equations 74–5
quintic formulae 74–5

R
radius 28, 124, 126, 128, 134, 139–41, 164, 170
rationalising the denominator 15–16
    with two terms 17–18
ratios 1–3, 10–11
rearranging formulae 28–30
rectangles 6, 8, 65, 189
reflections 114
remainder theorem 73–4
rhombus 126
roots 111–12

S
second derivative 170–2
sectors 126–8
segments 126–7, 126–8

simplification
    algebraic fractions 30–3
    expressions 4–6, 14–17
simultaneous equations
    intersection of two lines 93–5
    solving by substitution 68–9
    solving linear by elimination 69–73
    in two unknowns 67–73
sine (sin) 99–100, 104
    definition 107–8
    period 360° 109
    periodic nature 109
    $y = a \sin kx$ 114–19
sine (sin) graphs 108–9, 112–14
sine (sin) rule 128–32
    inversion 128–9
    in three dimensions 150–3
    use with the cosine rule 135–8
single points 144
spheres 138–9, 141
square roots
    and rearranging formulae 28
    simplifying expressions containing 14–17
squares 6, 8
stationary points 172–8, 184
stretches
    horizontal 114–15
    vertical 114
substitution, solving simultaneous equations by 68–9
surds, manipulation 14–18
surface area 14, 125, 138–41
symmetry 54–7, 62, 90, 199

T
tan (tangent) 99–100, 104
    definition 107–8
    $y = a \tan kx$ 114–19
tan (tangent) graphs 110, 112, 113–14
tangents 155, 158, 165–8, 173, 198
tetrahedrons 125, 153
three dimensions 142–3
    and the cosine rule 150–3
    lines and planes in 143–53
    Pythagoras's theorem in 149
    and the sine rule 150–3
    two dimensional representation 142, 143
transformations 114–15
trapezium 138, 189
triangles 8, 65
    and the angle between two planes 145–7
    area of 106, 122–6, 138
    and the cosine rule 132–5
    and the equation of a straight line 90–2
    equilateral 104, 106
    impossible 129
    and the intersection of two lines 96

isosceles 90, 103, 104, 124
and midpoints of a line joining two points 89
and Pythagoras' theorem 98–9
and Pythagoras' theorem in three dimensions 149
right-angled 101, 104
and the sine rule 128–31
and trigonometric functions for angles of any size 107–8
and trigonometry in two dimensions 99–107
use with the sine and cosine rule 135–8
trig ratios 104
trigonometric equations 110–14, 115–19
trigonometric functions 120
for angles of any size 107–8
trigonometric rules
cosine rule 132–8, 150–3
sine rule 128–32, 135–8, 150–3
trigonometry
real-world context 120
in two dimensions 99–107
true shape diagrams 143
two dimensional representation, of three-dimensional objects 142, 143

V

velocity 171, 172
vertex 54–7, 108, 133, 139–41, 148, 152
vertical stretches 114
volume 6, 14, 66, 138–41

X

x-axis, areas between a curve and the 189–91

Z

zero 59–60